REAL HISTORY

In *Real History: Reflections on Historical Practice*, Martin Bunzl charts a new direction for the philosophy of history. He proposes a synthesis between debates about objectivity among historians and recent philosophical arguments about realism. In his clear and direct style, Bunzl argues for an approach to history based on what historians actually do in contrast to what they say they are doing. Drawing on a broad literature including the works of Foucault, Geertz, Novick, Danto and Scott, the result is a new and exciting model for philosophy of history that casts objectivity and realism in a new light.

Martin Bunzl merges two parallel debates in history and philosophy. In his wide-ranging argument, he draws on relevant discussions ranging from: post-structuralism; to the philosophy of science; to the hermeneutic turn in anthropology; to debates about the history of women.

Real History is fascinating and essential reading for all those interested in a reconciliation of the debates about the methodology and study of the philosophy of history.

Martin Bunzl is a Professor of Philosophy at Rutgers University. He is also the author of *The Context of Explanation*.

PHILOSOPHICAL ISSUES IN SCIENCE

Edited by W. H. Newton-Smith
Balliol College, Oxford

THE RATIONAL AND THE SOCIAL
James Robert Brown

THE NATURE OF DISEASE
Lawrie Reznek

THE PHILOSOPHICAL DEFENCE OF PSYCHIATRY
Lawrie Reznek

*INFERENCE TO THE BEST EXPLANATION
Peter Lipton

*TIME, SPACE AND PHILOSOPHY
Christopher Ray

MATHEMATICS AND THE IMAGE OF REASON
Mary Tiles

METAPHYSICS OF CONSCIOUSNESS
William Seager

*THE LABORATORY OF THE MIND
James Robert Brown

*COLOUR VISION
Evan Thompson

*BRUTE SCIENCE
Hugh LaFollette and Niall Shanks

*LIVING IN A TECHNOLOGICAL CULTURE
Mary Tiles and Hans Oberdiek

*EVIL OR ILL?
Justifying the insanity defence
Lawrie Reznek

*REAL HISTORY
Reflections on historical practice
Martin Bunzl

*Also available in paperback

REAL HISTORY

Reflections on historical practice

Martin Bunzl

London and New York

First published 1997
by Routledge

11 New Fetter Lane, London EC4P 4EE
Simultaneously published in the USA and Canada
by Routledge
29 West 35th Street, New York, NY 10001

Typeset in Palatino by Routledge
Printed and bound in Great Britain by Creative Print and
Design (Wales) Ebbw Vale

British Library Cataloguing in Publication Data
A catalogue record for this book is available from the British
Library

Library of Congress Cataloguing in Publication Data
Bunzl, Martin.
Real history: reflections on historical practice / Martin
Bunzl.
p. cm. (Philosophical issues in science)
Includes bibliographical references (p.).
ISBN 0 415 15961 X (hc). ISBN 0 415 15962 8 (pbk.)
1. History–Philosophy. 2. Realism. 3. Objectivity. I. Title. II.
Series.
D16B935 1997 97 9101
901 dc21 CIP

ISBN 0–415–15961–X (hbk)
0–415–15962–8 (pbk)

I am seeking to rescue the poor stockinger, the Luddite
cropper, the "obsolete" hand-loom weaver, the "utopian"
artisan, and even the deluded follower of Joanna Southcott,
from the enormous condescension of posterity. Their crafts
and traditions may have been dying. Their hostility to the
new industrialism may have been backward-looking. Their
communitarian ideals may have been fantasies. Their
insurrectionary conspiracies may have been foolhardy. But
they lived through these times of acute social disturbance,
and we did not. Their aspirations were valid in terms of
their own experience . . . .

Edward Thompson, *The Making of the English Working Class*

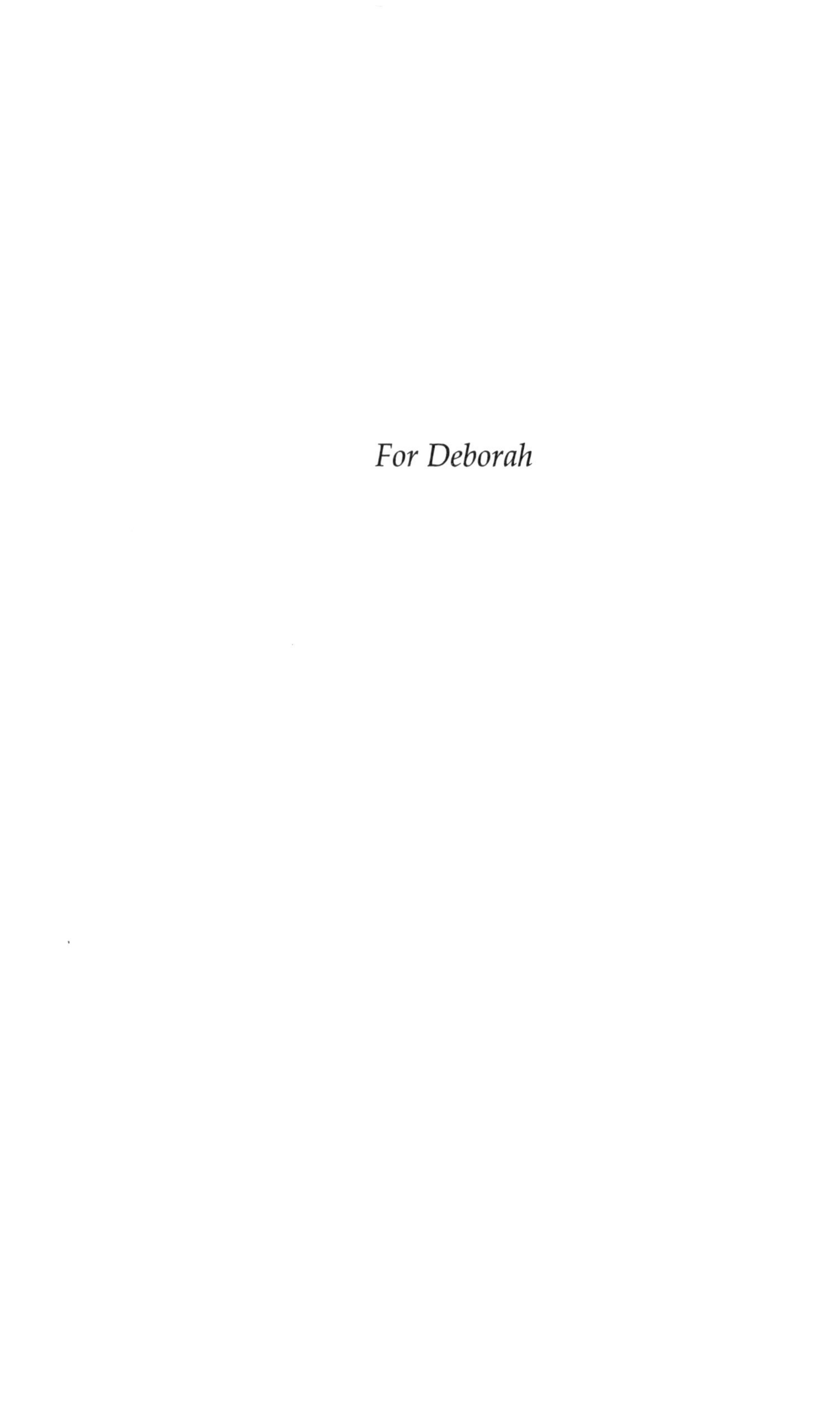

For Deborah

CONTENTS

ACKNOWLEDGMENTS

I am very grateful for the comments and criticisms provided by William Dowling, Deborah Hertz, Marion Kaplan, Donald Kelly, Peter Kivy, Peter Klein, Jackson Lears, and Laurent Stern, as well as William Newton-Smith, the editor of the series of which this book is a part. Thanks are also due to Miriam Rabinovitch for help in translating passages from French into English. This research was conducted with support from the Rutgers University Faculty Leave Program as well as a residency at the Blue Mountain Center in Blue Mountain Lake, New York.

Some of this material has appeared previously in a variety of articles; namely, "Archaeology without Excess," *Philosophical Forum*, 1995, vol. 27, pp. 27–36, "How to Change the Unchanging Past," *Clio*, 1996, vol. 25, pp. 181–193, "Meaning's Reach," *Journal for the Theory of Social Behavior*, 1994, vol. 24, pp. 267–280, "Non-cognitive Meaning Revisited," *Canberra Anthropology*, 1996, vol. 19, pp. 1–14, "Pragmatism to the Rescue?" *Journal for the History of Ideas*, 1995, vol. 20, pp. 675–680, "Scientific Abstraction and the Realist Impulse," *Philosophy of Science*, 1994, vol. 61, pp. 449–456, and "The Construction of History," *Journal of Women's History*, in press.

Finally, I should note that the title of this book takes its inspiration from *Real People* by Kathy Wilkes (Oxford, Clarendon Press, 1988)—readers of that fine book will readily see how my strategy here about history has been influenced by her work.

Martin Bunzl
Rutgers University

INTRODUCTION

The practice of history has to contend with philosophies of history that exist in two different worlds. One is inhabited by philosophers, the other by historians. Workers in these different worlds rarely attend to the work of each other, despite overlapping interests. Nowhere is this more true than in questions surrounding what philosophers call realism and what historians call objectivity. The sorry tale of philosophers' irrelevance is documented in Peter Novick's study of the history of debates about objectivism between historians in the United States over the last one hundred years.[1]

In this book I attempt to construct a synthesis between these traditions. Such a synthesis can be bruited with two different levels of ambition. At the narrow level we can try to integrate the philosophy of history as written by philosophers and by historians. But trying to do so in this way produces an artifact that hangs in isolation both from other areas of philosophy and, more importantly, from historical practice as well. Attempting a synthesis of the traditions within these broader contexts is much more interesting. And that is what I aim to do. But in doing so, I restrict myself to what historians call "The Objectivity Question."[2]

Novick nicely summarizes a consensus position among historians about how to think of objectivity, including its assumptions:

> The assumptions on which it rests include a commitment to the reality of the past, and to truth as correspondence to that reality; a sharp separation between knower and known, between fact and value, and, above all, between

> history and fiction. Historical facts are seen as prior to and independent of interpretation: the value of an interpretation is judged by how well it accounts for the facts; if contradicted by the facts, it must be abandoned. Truth is one, not perspectival. Whatever patterns exist in history are "found," not "made." Though successive generations of historians might, as their perspectives shifted, attribute different significance to events in the past, the meaning of those events was unchanging.[3]

The philosophical sentiments that underlie these views are not much in fashion in current historiographic writing, which is firmly anti-realist under the influence of post-structuralism.[4] Yet they form the core of this book. My aim is not a defense of these philosophical sentiments but a re-examination—a re-examination that proceeds against the background of historical practice. My thesis is rather modest. It is this: it is hard not to be an "objectivist" in practice. But the kind of objectivist you end up being does not embrace all of the aspects of objectivity in the quotation above. Separating the wheat from the chaff is the intellectual work of this project.

For the past twenty years the philosophy of history (as written by philosophers) has become largely disengaged from the central concerns of the philosophy of science and the philosophy of language. Yet in both the philosophy of science and the philosophy of language one finds striking parallels to some of the Parisian influence on recent work by historians writing in the philosophy of history. Non-philosophers often think of analytic philosophy and positivism as being synonymous. Yet within (mainline) analytic philosophy, positivism was overthrown as the dominant philosophy more than forty years ago, and, long before Rorty, it was reasonable to characterize a (if not the) dominant view of Anglo–American philosophy as being composed of three tenets: holism as a theory of meaning, coherentism as a theory of truth and (more recently) anti-realism as a view of metaphysics.[5] Furthermore, as in the case of post-structuralists, these positions derive directly from considerations of language.

If one begins to construct a philosophy of history based on these views, the road runs nearly parallel to the French highway. In fact, in crucial respects, I think both roads run parallel. As such, criticisms of one program ought to apply to the other.

Accordingly, recent *challenges* to the dominance of the post-war program in analytic philosophy may be of value in the debate about the philosophy of history. In particular, there is a movement in analytic philosophy of science which is anti-theoretical, to the extent that it urges against imposing on science any particular view with regard to ontology "from the outside."[6] The upshot of this view is not that history has no ontological commitments—they are legion. But we should examine them by working from the inside out. A consequence of this view is that such an analysis may yield quite heterogeneous results depending on the particular historiographical context. Hayden White would have it that all history has within it an implicit philosophy of history.[7] I think that we have to worry that the practice of history may underdetermine the philosophy of history. That is to say, we have to worry that practice may not be capable of providing the brute data to decide between competing philosophies. Nonetheless, I think that the practice of history does allow us to decide some competing claims—if not between whole philosophies.

If we were to accept Novick's characterization of historical objectivity as a starting point, we could examine counter-positions by defining them by what it is about historical objectivity that they reject. But there are reasons not to accept Novick's view—even if it is descriptively accurate as an account of the concept of objectivity as historians have used the term.

One of the great virtues of *That Noble Dream* is Novick's account of the "Rankean" dictum that the aim of the historian should be to capture the past as it actually was.[8] I refer to the dictum as "Rankean" rather than Rankean deliberately. For, as Novick's account documents, the Americanization of Ranke's views transformed them by stripping away their Hegelian underpinnings.[9] In *That Noble Dream*, Novick provides us with a history of how "Ranke's epistemology was naturalized into an English empiricist idiom."[10] The result is a presentation of how American historians have conceived of the notion of objectivity. Novick's treatment of how this conception has been used within the American historical profession is exhaustive,[11] and against the backdrop of that exhaustiveness it becomes all the more obvious just how little *philosophical* aspects of the question of objectivity in history have played a role in the last one hundred years of the profession. Not only have philosophers been irrelevant but so has philosophy. For what Novick's history makes so clear is the way in which a conception of historical objectivity has provided a stalking horse for a variety of other issues, among which one was paramount.

The sweep of *That Noble Dream* provides an opportunity to see the use of the canon of objectivity against a variety of alleged counter-conceptions—including the conception of history as informed by moral judgments and the conception of history as informed by political interests. But the real driving force that gives unity to Novick's construal of this history is the relationship between objectivity and the aspiration to achieve univocality in the claims of scholarship, and, perhaps not surprisingly, as Novick's account makes clear, that aspiration derived from something less pure than a philosophical interest. The drive for historians to speak of the past with one voice was a powerful tool in the nineteenth-century movement of professionalization. It also became a central pillar of the rhetoric used in the attempt to resist the increasing heterogeneity of the profession in the middle of the twentieth century. But, notwithstanding this central interest, one of the philosophical opportunities provided by Novick's work is his treatment of the status of historical facts.[12]

Both Beard and Becker are often thought of as the pre-eminent *American* anti-realists of the first part of this century.[13] It is to Becker that Novick (correctly) attributes the position that "No historical synthesis could be 'true' except 'relative to the age which fashioned it.' "[14] And, in general, Novick's thesis is that it is correct to attribute relativism to both Becker and Beard. But it is important to notice just how circumscribed this relativism was intended to be. To do so we need to distinguish between relativism about facts and relativism about grander historical constructs. Novick writes that:

> there were no professional American historians, unguarded and casual remarks aside, who denied that at least singular factual statements about the past could be "true" for practical purposes. There was, however, considerable "skepticism" about the likelihood, or possibility, of creating a cumulative, convergent, "corresponding" picture of significant events and epochs of the past.[15]

For both Becker and Beard the facts in themselves were not in question on Novick's account[16]—nor the empirical method as the route to those facts. The "problems" arose only if one moved beyond the facts—to synthesis; and Becker and Beard counted themselves among those who argued that such history was the only history to be had.[17]

What gave "the facts" this privileged status and insulated them from the force of relativism? The most plausible contemporaneous philosophical reason to question the status of facts about the past would have been based on positivist considerations: such facts are not observable. But Becker and Beard were not driven primarily by philosophical considerations. Their concerns were much more attuned to purely internal debates about the function and role of history and its relationship to presentist interests, and to the extent that they *were* influenced by wider methodological issues of their time, it was pragmatism not positivism to which they were drawn.[18] (As Don Kelly has pointed out,[19] the philosophical underdevelopment of this American debate ought to be contrasted with what was a contemporaneous debate in Germany. If there is one thing that offers a key to the difference, it is the pervasive influence of neo-Kantianism, in all of its variants, that formed a background to the German debate.)

As long as you allow for the realist status of facts, then the case for anti-realism rests at minimum on making the argument that history cannot be composed of just such facts. And while Novick is perfectly aware that one can attack such a conception of facts, his reconstruction of debates about realism within the profession rests heavily on the idea that such debates left the status of facts themselves undisturbed. For all the confusion in the positions of both Becker and Beard that Novick points to, I think it is fair to say that Beard thought a history of just the facts was not just uninteresting but in fact impossible. Beard thought so because he embraced the idea that, even though history includes facts, it also involves an ineliminable *contemporary* component in the organization and selection of "the facts":

> It [history] is, as Croce says, contemporary thought about the past. History as past actuality includes, to be sure, all that has been done, said, felt, and thought by human beings on this planet since humanity began its long career. History as record embraces the monuments, documents, and

> symbols which provide such knowledge as we have or can find respecting past actuality. But it is history as thought, not as actuality, record, or specific knowledge, that is really meant when the term is used in its widest and most general significance.[20]

While for Becker:

> The facts of history are already set forth, implicitly, in the sources; and the historian who could restate without reshaping them would, by submerging and suffocating the mind in diffuse existence, accomplish the superfluous task of depriving human experience of all significance. Left to themselves, the facts do not speak; left to themselves they do not exist, not really, since for all *practical* purposes, there is no fact until some one affirms it.[21]

However we cash in the details behind the idea of "significance" in both of these passages, the core, familiar idea is of course that there is no history without interpretation, and it is with interpretation that anti-realism comes into the picture. The debate, as it were, does not get going until we reach interpretation. Now, in principle, the realist has two options: to argue for history without interpretation or to argue for a realist interpretation of interpretation. But, on Novick's account, it is only the first of these two options that gets seriously pursued. As such, the objectivity question becomes equivalent to the question of the eliminability of interpretation in history.

But we pay a high price for allowing this collapse of the question of objectivity into the question of interpretation. The philosophical difficulties with this way of forming up the debate are two-fold. On the one hand, *if* facts are accorded an unproblematic status, then it is not obvious, *prima facie,* that interpretation (or at least some kinds of interpretation) cannot be given the same status. That is to say, it is not obvious that the "standard" view of facts is not robust enough to support a very minimal account of interpretation—minimal enough to be compatible with a realist view of history. On the other hand, if facts are *not* accorded an unproblematic status, then of course the debate about realism will be engaged right away, and, as we will see, there are both philosophers and historians who have done just this. But beyond the philosophical, we pay an historical price as well; it is just not clear that cutting the world into the factual and the interpretative can do justice to the enterprise of understanding the past—including past historiography. For to do so forces us to see historians with interpretative interests as anti-objectivists. To avoid this we first need to disentangle the question of objectivity from the question of interpretation. That task takes up the next chapter. Once we have done that, we can turn back to history.

1

OBJECTIVITY RECONFIGURED

Let us begin again with Novick on objectivity:

> The assumptions on which it rests include a commitment to the reality of the past, and to truth as correspondence to that reality; a sharp separation between knower and known, between fact and value, and, above all, between history and fiction. Historical facts are seen as prior to and independent of interpretation: the value of an interpretation is judged by how well it accounts for the facts; if contradicted by the facts, it must be abandoned. Truth is one, not perspectival. Whatever patterns exist in history are "found," not "made." Though successive generations of historians might, as their perspectives shifted, attribute different significance to events in the past, the meaning of those events was unchanging.[1]

And, for the purposes of argument, let us accept this as a consensus position among historians about how to characterize objectivity.[2] Even so, from a philosophical point of view, we need to ask whether this is the only way or the best way to do so. I take the descriptive position Novick outlines to include a number of elements that it shares in common with the philosophical position of metaphysical realism as it is often characterized. On that characterization, metaphysical realism is the position that there exists a mind-independent world, claims about which are true in virtue of their correspondence with features of that world.

OBJECTIVITY AS METAPHYSICAL REALISM

Now metaphysical realism itself is a position that has been out of favor in the philosophical literature for some time, but, before we examine this literature, we need to ask if Novick's characterization goes beyond metaphysical realism. In order to do so, it will be useful to break down the above quotation into its constituent parts. I take these to be as follows:

1 There is a fact of the matter about the past that is settled by the correspondence of historical accounts with the past.
2 Interpretation is secondary to facts and facts always trump interpretations.
3 Truth is not position-relative.
4 If history has patterns, they are found and not made.
5 The meaning of history is unchanging.

And I take the issue to be whether "objectivity" needs to assume anything beyond point 1, for I take point 1 to be essentially equivalent to the position of metaphysical realism as outlined. So what do the other points add? Strictly speaking, point 2 does not assert that interpretation is inconsistent with objectivity, only that, as it were, the facts come first. Thus the value of an interpretation may be judged by "how well it accounts for the facts," but that is not to say it is the *only* basis for valuing an interpretation. I say "strictly speaking" because the spirit of *That Noble Dream* is much more in line with reading point 2 as the assertion that accounting for the facts is the *only* basis for judging competing interpretations. But whichever of these alternatives you embrace makes little difference here, for, on either view, point 2 asserts the primacy of facts. And it is by way of point 1 that an account of facts gets offered implicitly; namely, that facts are truths that get settled by their correspondence to the "real" past. So on the argument I have just given, I take point 2 simply to be an extension of point 1. Whatever the role of interpretation, only the facts matter when it comes to the question of objectivity.

What about point 3? What does it mean to say that "truth is one" and not "perspectival"? In the context of historiographical discussion, I take this to be anti-presentist. That is, it stands against the view that, unlike the past itself, our historical writing about the past is inextricably tied up with considerations not in the past itself. That might be thought to come down to this:

correspondence with the past is not the only relevant consideration in evaluating historical writing (and, as the "present" changes, so too will its histories). Still, so far, truth itself has not entered into these presentist considerations. Point 3 is consistent with considerations from the "present" being relevant. It just rules out their relevance as far as truth-considerations are concerned. So here, too, point 3 turns out to be parasitic on point 1 in asserting that, as far as truth is concerned, the past is all that is relevant.

Point 4 is a conditional whose antecedent few today will think is satisfied at any sort of grand level (of, say, laws). But irrespective of the level at which we engage the question of "patterns" in history, we would do well to think of a parallel in science: just as you can be a realist in science about unobservable entities without being one about laws, so too in history you could be a realist about the events[3] in the past but not about "patterns" about the past. As such, point 4 is a point about the scope or range of point 1; namely, whether it applies to patterns about the past.

Finally, what about point 5? Here the situation is complicated by the use of the term "meaning." If you allow that the significance of the past may change for differently situated actors in the present, in what sense should we speak of the "meaning" of past events as unchanging? "Meaning" means a variety of things, none of which is synonymous with "truth." Nonetheless, I take this point to reinforce the principle that, even as our interests may change with respect to the past, that does not change those events or their meaning which is rooted in the "reality" of the past.

Based on the above considerations, on Novick's account, what is crucial about the assumptions of objectivity is a version of metaphysical realism applied to the past. But what is the standing of metaphysical realism itself as a position? Historians often object to versions of what is essentially metaphysical realism on what are epistemological grounds. But it is important to note that the position (in contrast to scientific realism) makes no epistemological claims *per se*.[4]

METAPHYSICAL REALISM

When historians deny that there *exists* a past to which history can be said to correspond, an inference is often made to the conclusion that it therefore follows that history is somehow "constructed" *tout court*. But the fact that what happened in the past no longer exists does not in itself undermine the notion of correspondence for history any more than it does in the area of cosmology or the area of criminology. To claim that it does is to assume erroneously that correspondence can only obtain in the here and now. That it need not makes for an epistemological problem. But from this it does not follow that there is an *ontological* problem about the past. Correspondence may be hard to establish epistemologically without it being ontologically problematic.[5] That is to say, unless you are a skeptic, the fact that our epistemic situation forces us to make inferences from evidence does not undermine the ontological conclusions of what it is we infer from that evidence.

Whether we have the means to discover features of a mind-independent world may be treated as an open epistemological question for the metaphysical realist. Nor need there be an assumption that we could not turn out to be wrong in our knowledge claims.[6] Still, that is not to say that the truth of such claims is not (to use Jamesean language) "absolute and true always" for a metaphysical realist—but both of these features are much weaker than this (Jamesean) rhetoric implies, in that they are conditional on what we take to be true actually being true. If it *is* true that Nixon was the 37th president of the United States, then that is a truth that is always and absolute. Of course, it may turn out that it is *not* true—we can speculate why: maybe (to vary a story of Putnam's) the *real* Nixon was replaced by a cleverly disguised Martian robot after he lost to Kennedy, but we won't find this out for another few years until his body happens to be exhumed for study. And if, in fact, it is *not* true, then it never was true—always and absolutely . . . instead, we were just mistaken in our belief about the 37th president of the United States.[7]

I don't think the difficulty of metaphysical realism is one of arrogance about the veridicality of our beliefs about the world—metaphysical realism is compatible with a variety of epistemic stances, from the overconfident to the overcautious.

Nor is the difficulty one of the notion of truth itself. For, even though metaphysical realism is naturally allied with a correspondence notion of truth, it need not be. Indeed, I think we can proceed just as well in an account of metaphysical realism without truth itself. For any statement you make that uses the word truth assertorily, I can issue an equivalent one that dispenses with the word. You say "It is true that Richard Nixon was the 37th president of the United States," to which I say "Richard Nixon was the 37th president of the United States." Why is this not philosophical sleight of hand? Because I don't want to have to define "truth." What about the assertion "Richard Nixon was the 37th president of the United States"? *Maybe you will say you want to know if it is true.* But then most likely (to steal an image from Appleby, Hunt and Jacob)[8] we will be off to the archives—we will be involved in an evidential debate. We have a common enough discourse for us both to take it that the sentence "Richard Nixon was the 37th president of the United States" is true just in case Richard Nixon was the 37th president of the United States. Now you can cash this in via a correspondence theory of truth if you want to, but nothing prevents you cashing it in by another theory or simply leaving it as is.[9]

Rather than a matter of epistemology or a theory of truth, the core difficulty that I think metaphysical realism faces is instead a problem of reference. If we allow that human knowledge is articulated via language (of course there are other senses of knowledge, but they are not at issue here), the challenge the metaphysical realist faces is just how we manage to break out of language to (as it were) connect with or accurately to reflect the mind-independent world. The problem is not one of a misplaced desire exhaustively to reflect that world—just worrying about the statement "Nixon was the 37th president of the United States" raises the problem, and here the worries even start with the most straightforward part of that statement. How does "Nixon" pick out Nixon?[10] Now it may seem that this worry about reference for accounts of metaphysical realism will only arise if we take its polar opposite seriously—that is, if we take seriously the non-referential views about language championed by post-structuralists. And that would create a problem for the proponent of a middle position who wanted to reject *both* metaphysical realism and post-structuralism. But it is not necessary to embrace the linguistic extravagance of a Derrida in order to make stark the referential challenge that the metaphysical realist faces—the indeterminacy of translation does just as much damage.[11] While it is easy to reject a Derridean view of language because of the very breadth of its clash with nearly every important feature of our linguistic performance, I do not think the thesis of the indeterminacy of translation can be dismissed nearly so easily.[12]

Let us outline the moves open to the metaphysical realist under this line of attack. One alternative would be to reject the indeterminacy arguments themselves. A second alternative would be to modify metaphysical realism to make it a thesis that does not depend on reference at all. But neither of these turns out to be very easy to accomplish. In what follows I will do no more than outline the difficulties each approach faces.

Nearly all defenders of contemporary realism have taken the first option. But to do so persuasively requires some outline of just how determinative reference might take place. The favored way of answering this challenge has been to embrace some sort of causal theory of reference. That is, a theory in which a causal link may be said to connect referential acts to referents. So, for example, as Kripke argues, names may be thought of as connected to people so named by way of their initial naming (their "baptism"), to which all other successful uses are directly or indirectly linked.[13] However, in such accounts, not any causal link will do. All causal theories fall prey to counter-examples of the wrong *kind* of causal connection.[14] To work, such accounts need something more to distinguish between right and wrong kinds of causal account. But that has proved to be exceedingly hard to do while holding on to the virtues that such accounts possess over their competitors.

The challenges to a causal theory of reference suggest that, if we could sidestep the referential component of metaphysical realism, we could avoid these problems altogether, and that is a strategy that has been defended most vigorously by Michael Devitt.[15] But such a view is hard to make sense of if (abstract) theories are to fall within the subject matter of metaphysical realism, as I think they do if we take the aspirations of the doctrine at face value.

REALISM WITHOUT METAPHYSICS

What if we proceed without metaphysical realism in the light of these challenges? Then the road quickly divides between embracing some form of full-blown anti-realism or the hope of some sort of middle position between metaphysical realism and anti-realism.

Now it seems to me that, all other things being equal, anti-realism is in a much better position than any version of realism from a philosophical point of view. Because realism is a much more ontologically laden position than anti-realism, the burden of proof will always be on realism and, by extension, on any middle position, and it remains to be seen whether middle positions can meet this challenge. Certainly the most obvious place to look is in the work of Hilary Putnam.[16] Putnam's account of "internal realism" is driven by the commitment to avoid any assumptions about the success with which language may be treated as determinant. The reference of all terms is to be allowed to be radically indeterminate. And that is to include terms like "truth," "objects," "events," "facts," and even "reference" itself. Thus even our attempt to talk about these matters is subject to these indeterminacies. Hence there can be no talk about realism except in so far as it is relativized to an interpretation, and that frame of interpretation will be only one of an infinite number of alternative interpretations. Hence the notion that such a version of realism is "internal." But how does such a position avoid collapsing into full-blown relativism and, with it, full-blown anti-realism? Putnam is aware of this challenge, but sidesteps it by arguing that it arises out of a misconceived idea that a "totalistic explanation" is available to us.[17] *All* we have available to us is our internal explanations. Hand in hand with that goes the (Wittgensteinian) idea that it is a philosophical pretension to think we can move beyond the features of everyday use in seeking to analyze the central terms that drive the realism debate.[18] Suppose we accept such an account for the purposes of argument and "go" internal. Still, even from such an internal perspective, what is to prevent me asking if our stance is realist or anti-realist or something in between? Nothing makes such a question incoherent—even if this version of realism is more tempered than the full-test version of metaphysical realism with which we began.[19]

But can such a position withstand scrutiny?

TWO MODELS

In a series of important papers, Arthur Fine has developed and defended the view that the proper reading of scientific *practice* is neither realist nor anti-realist.[20] Instead, he argues that realism and anti-realism both add something extra to a core position *which is neither*.[21]

For Fine, that core position involves treating science as of a kind with common, garden-varieties of knowledge. That is to say, "true" in science ought to be taken as having no other connotation than it does in ordinary life. What does this amount to?

> What is it to accept the evidence of one's senses and, *in the same way*, to accept confirmed scientific theories? It is to take them into one's life as true, with all that implies concerning adjusting one's behavior, practical and theoretical, to accommodate to these truths.[22]

Fine thinks this is where we should stop *without an elaboration of the concept of truth at work here*.[23] He suggests instead that, viewing science in historical context and as a human, social process, we take the scientist as operating with the same conception of truth that characterizes other human concerns.[24]

For Fine, the core position is adequate to an account of science. Rather than thinking of realism and anti-realism as full-blown alternatives to this core position, Fine thinks of them as each adding something (illicit) by "*inflating*" the core position motivated by misguided hermeneutic commitments to "the global enterprise of 'making sense of science'."[25]

Both Alan Musgrave[26] and Richard Miller[27] have raised a technical objection to Fine's approach by arguing that its cogency hinges on how the notion of truth is understood. This objection comes down to the claim that, to the extent that Fine's position makes use of an ordinary conception of truth, it makes use of a notion of correspondence and hence is in effect a realist doctrine. I am unmoved by this line of criticism because, as I have already indicated, I do not think that truth and correspondence have to go hand in hand even on an ordinary conception of "truth," and I do not think that realism has to assume a correspondence theory of truth.

Furthermore, the criticism notwithstanding, as Richard Miller has argued,[28] there is a level at which we can approach Fine's argument that sidesteps this issue. What Fine calls "The Natural Ontological Attitude" (NOA) is one that resists the idea of moving outside science in an attempt to provide a global characterization of it:

> NOA thinks of science as an historical entity, growing and changing under various internal and external pressures. Such an entity can be usefully studied in a variety of ways, sociological, historical economic, moral, and methodological—to name a few. One can ask a variety of questions about particular developments in particular historical periods. Sometimes there will be a basis in the practice itself for answering such questions. Sometimes that basis will support several plausible answers. Sometimes it will be clear that there is no basis in science itself for addressing the question, and then one must judge for oneself whether it is worth adding attachments on to science so as to make a place for the question, or whether one should just let the issue drop.[29]

For Fine, the error of both realists and anti-realists is their failure to appreciate the characteristics of science as a social enterprise, and Miller argues convincingly that, whatever the weakness of the way in which Fine sets up the contrast between the NOA and both realism and anti-realism, the central issue is that both feed on "scaffolding external to science."[30]

Of course there is "external" and "external" and, as Miller points out, depending on what you mean, Fine's account turns out to be either conservative in insulating scientific practice from criticism *or* implausible in its characterization of realism, including the idea that it is imperialist about the goals of science.[31] Though Miller doesn't put it quite this way, what is also at issue is what I will call the globalism that Fine wants to assign to both realists and anti-realists. That is to say, the realism and anti-realism under discussion are to be understood as competing claims to be applied uniformly to science as a whole rather than as piecemeal theories about particular scientific theories. Indeed, such a view of the global reach of both realism and anti-realism is not unreasonable, given how debates about realism have progressed in the philosophy of science over the past twenty-five years. Yet it is a crucial view that turns out to be central in allowing Fine to distinguish the NOA from its competitors.

Whether or not you take issue with the way in which Fine sets up the distinction between the core position and its additions (both realist and anti-realist), his argument rests heavily on the distinction between the NOA as internal to science and its opponents as external. But what makes realism and anti-realism external? In fact, Fine gives no argument to exclude realism and anti-realism as *internal*. Still, what does rule this out as plausible is the implicit treatment of realism and anti-realism as global theories. For, if we follow Fine in treating "science as an historical entity, growing and changing under various internal and external pressures,"[32] such (internal) globalism, while *possible*, is certainly not necessary nor particularly likely. For the core of such an historical view of science is its very openness to the potential for change, which is just the feature that renders *any* global claims we might make about the nature of science suspect.

If globalism is set as a characteristic of realism and anti-realism, then the implausibility of globalism will infect the plausibility of claims about realism and anti-realism themselves. They will stand or fall with their global nature. But suppose we renounce *a priori* globalism. Then in fact realism and anti-realism begin to look like plausible internal candidates along with the NOA. For suppose we agree that "Such an entity [i.e. science itself] can be usefully studied in a variety of ways, sociological, historical, economic, moral, and methodological—to name a few."[33] Then why can we not include questions of realism and anti-realism on this list, as long as we remember that we are asking "questions about particular developments in particular historical periods?"[34]

If we can, then the issue becomes whether or not "there will be a basis in the practice itself for answering such questions." Can practice ever yield us up an answer? If we allow for non-global versions of realism and anti-realism, no principled argument has been provided to show that practice alone could not support one or the other under particular circumstances. And a principled argument is what we need. For, by allowing for the historical variability of science as a social enterprise, the claim that past practice has not supported realism or anti-realism does not exclude its possibility in the future.

To exclude it in the future we need to do something more.

Short of a principled argument that would rule out realism and anti-realism under any circumstances, a sensible strategy would be to place limits on the range of those circumstances. Such a strategy suggests itself from the work of Howard Stein.[35] Like Fine, Stein thinks that both realism and anti-realism (in the form of instrumentalism) are irrelevant to scientific practice, and, in one sense, like Fine, Stein thinks they have a common core. But Stein's approach is different from Fine's in two respects. First, Fine casts the core position in terms of an unanalyzed conception of truth. On the other hand, Stein wants to eschew talk of truth altogether in the core position. His core commitment for a theory is just that theories of phenomena "afford . . . a representation [that is] . . . in a suitable sense correct, and in a suitable sense adequate . . . ."[36]

Like Fine, Stein thinks that if instrumentalism is taken to be nothing more than predictive adequacy, it is too narrow. But given his construal of the "core," Stein thinks it is easy to provide an account of instrumentalism that is as broad as we like. Whatever our interests, we simply render them as the content of our *instrumentalist* interests. Think of the instrumentalist as treating theory as an instrument and no more. Then that instrument can be thought of as being put to a variety of different uses. One of those uses can be prediction, but there can be others. So anything we do with theories can be assimilated into this framework.[37]

But what about realism? The (traditionally conceived) realist wants to argue that judging theories to be correct and adequate in a suitable sense is not enough. Like Fine, Stein thinks this realist asks us to do the epistemically impossible—for, beyond representing phenomena correctly and adequately, "how in the world could we ever tell what the actual case is?"[38] But what is left of realism if we give up such talk? Stein thinks the position becomes not importantly different from the expanded version of instrumentalism. *Both* are guided by theories judged by their correctness and adequacy as representations in the "suitable sense."

So here we have a core position for scientific practice that supports *both* realism and anti-realism (in the guise of instrumentalism)—at least when they are suitably redefined. If that is the case, then practice will *not* yield data to favor one over the other. And if the argument can be applied to history, realism and anti-realism will then only constitute a distinction in a philosophical meta-realm disconnected from the dictates of historical practice.

However, if Fine's core position had to be defended against the criticism (of Miller and Musgrave) that it is too easy for realists to accept and too hard for anti-realists, in Stein's case things are the other way around. Stein paints the realist as one who wants to *add* a condition on what is in a "suitable sense correct, and in a suitable sense adequate" for a theory to represent phenomena. But this is not quite right. Suppose we treat this as a matter of fitting the data. For a realist, what counts is not an interest *above* and *beyond* an interest in fitting the data. For the interests that drive realism render fitting the data as merely a means to an end—indeed, one that can sometimes be sacrificed when one of two competing theories exhibits significantly greater explanatory power while the other exhibits greater empirical adequacy. (Nothing hangs on this particular choice of how to instantiate the notions of correctness and adequacy here.) The same argument can be raised for any substitute condition that falls short of full-blown realism. As such, the real fight between realists and anti-realists (even relaxed and reasonable ones) has to do with what counts as the core of the scientific enterprise. That is to say, there is no guarantee of a common core between the realist and Stein's anti-realist.

Put that way, we do not seem to be much better off than before. We lack a general argument to show the irrelevance of realism and anti-realism. Yet, though Stein's account may fail in its formulation, like Fine, he has a general story to tell about science that is more important. That story revolves around the centrality of mathematics. Over time:

> what tends to persist . . . [is] not the features most conspicuous in referential semantics: the substances or "entities" and their own "basic" properties and relations, but the more abstract mathematical forms.[39]

If you think of science in this way, success becomes the development of a mathematical account which the world may be said to instantiate. That may make it seem like an opportunity merely to relocate the debate between realists and anti-realists. Now the issue becomes how to understand "instantiation." But at the same time in this locus it is much easier to be indifferent between

Stein's versions of realism and instrumentalism. For, if we think of the canon of correctness and adequacy, what are the choices for a "suitable sense" for these when it comes to the mathematical accounts? In response, the realist wants to say "truth!" coupled with correspondence. But what counts as correspondence in this context? It is not that one cannot answer this question. But the answers will involve entities whose own ontological status is highly problematic. Contrast this with the idea of realism applied to the garden-variety furniture of the universe. Of course the committed anti-realist will claim that this distinction relies on nothing more than a prejudice about the furniture of the universe. And it is at this juncture that realism and anti-realism seem to clash most naturally. But I think we should resist the temptation to let the argument settle in at this point. For this is precisely the point at which the debate between realism and anti-realism becomes a debate about differing interpretations of the correspondence theory of truth. Yet concentrating on this debate misses what it is that really cathects the issue. Instead, I think that realism is properly understood as a style of reasoning[40] that only engages in certain kinds of situations. The paradigms of these situations involve scientific accounts of macroscopic objects and such accounts are judged as adequate and correct to the extent that they coincide with features of the world that we take for granted. As such, the "correspondence" here is not between our theory and the world but between our theory and features of the world *that we take for granted*. Nor is realism, on this view, a position that falls out of a theory, but rather it is a mode of reasoning that we bring to bear in certain kinds of model building for which we already have antecedent realist commitments to the subject independent of our model. And such a mode of reasoning is hard to engage for mathematical accounts.

But should we accept Stein's view about the centrality of mathematics? On the above argument, to do so means underplaying the significance of our questions about things like events. That seems to me to be quite right, and right prospectively with respect to science's future as well as its past. As such it holds the potential for a general account of why the debate between realism and anti-realism is not very important given our scientific interests, even though, as we saw in my treatment of Fine's account, we cannot rule out the possibility that other interests might come to the fore that would undermine this conclusion. On the argument so far, all they would have to be is non-mathematical. But in fact I do not think that the issue is simply a counterpoint between the level of concrete particulars and mathematics. Rather, what is involved is our scientific interests in abstraction understood more generally. For it is easy to see how the same problem of how to understand the distinction between realism and anti-realism arises in non-mathematical contexts of abstraction as well.

Nowhere is this more true than in biology. To understand evolutionary history is to understand the machinery of just how it was that some biological attributes were disfavored over others. Here, realism comes in quite straightforwardly, in that the subject matter does not demand such a high degree of abstraction as to prevent us from assimilating it to our everyday style of reasoning. Yet, in and of itself, evolutionary history is not very interesting. Unlike evolutionary history, when it comes to evolutionary theory, the interesting questions of biology quickly reach beyond the level of the particular. And given the variability of the particulars in the biological world, reaching beyond the particular quickly forces us to generalize over that variability. The models that result are no longer models of the world but models of very narrowly circumscribed *abstracted* properties of the world. The result is that the idea of *correspondence* as a one-to-one relation between features of the world that we take for granted and features of our theories loses most of its meaning. Thus, in the case of natural selection, at the level of individual reproductive units, the details of historical circumstances are exquisitely variable—so much so that there can be no generalization at this level of description. To generalize we have to recharacterize what goes on (in the account of fitness) as a way to subsume this variability. Thus, what makes for fitness can vary radically, and the resulting theory, even though it names a real property, produces no straightforward instantiations of fitness at the level of individual units.[41]

BACK TO THE PAST

The example of biology just discussed broached the distinction between evolutionary history and evolutionary theory. The upshot was that only at the level of theory does the abstraction of science undermine the distinction between realism and anti-realism. Can we learn anything from this argument about the case of history itself?

In a derivative sense there is an obvious answer to this question. To the extent that historical studies make use of theories from other disciplines and those theories have the characteristics that we have been examining, then so too will the historical studies. But what about on its own accord? To learn anything from the case of biology and the division of evolutionary history from evolutionary theory, we need to ask whether a similar division obtains in historical work itself.

At first blush it may seem as if it takes very little theorizing in history to force a level of discourse in which the notions that drive realist intuitions get undermined. Indeed, the more varied the underbrush of day-to-day life about the past is, the more quickly this supervening level of theory will be reached. It is the variability of causal mechanisms that yield reproductive advantage that forces the introduction of a theory that makes use of supervening categories in biology. In physics, you can go much further up the ladder of theoretical complexity before you run into the problem, because of the underlying uniformity of the substrate. Particles are much more alike than the range of reproductive phenomena and, in this sense, many of the subjects of historical study are even more varied than those of biology. Hence any attempt to generalize *over* them will quickly undermine the hope of any straightforward notion of correspondence (between our theory and features of the world that we take for granted). Hence, at least on the argument so far, it seems unlikely that realism as a style of reasoning will enter into history.

But I think that there are two flaws that underlie this line of reasoning. First, it assumes that history requires generalization. And second, it assumes that any such generalization requires abstraction. Let us avoid any false bravado in speaking of history as a whole—allowing that it may be composed of a variety of interests and methodologies. Still, I want to assert of at least some parts of the enterprise that historical research does not involve much generalization. Moreover, I want to claim that where such generalization does occur, it involves much less abstraction as a precondition than what we see in biology. The reason why is quite straightforward: where generalization does occur in history, it is, by and large, small in scope. The smaller its scope, the lower the level of generalization, and the lower the level of generalization, the greater the chance of covering things of the same kind without need of abstraction, even in a variegated world.

But even if (some) history involves no generalization, or where it does, it involves only limited abstraction, these do not in themselves constitute an affirmative argument for a role for realism as a style of reasoning in historical research.

How might such an argument be constructed?

Suppose, echoing Fine, we start with historical practices, and ask the following question: what are the minimum conditions of adequacy needed if we are to make sense of them? Two caveats: first, there need be no unitary answer to this question given that there need be no unitary practice when it comes to history; second, the question is posed as a conditional. One can always choose to reject the antecedent of the conditional and conclude that history makes no sense. Let me illustrate what I have in mind with one condition of adequacy that (perhaps not all but at least many) accounts of history must satisfy: *accounts of history have to make sense of the notion of disagreement among practitioners.* As we will see in more detail in Chapter Three, we can get a lot of philosophical mileage from such a modest claim. Merely by allowing that disagreement is possible, you thereby commit yourself to the notion that disagreeing historians are talking about the same thing. Disagreement assumes some area of overlap about which the disagreement takes place.

Be that as it may, the condition of adequacy just posed does not force an account of history that faces outward, as it were. And this leaves open what is most at stake—namely whether historical practice demands a realist interpretation. But I want to suggest that, for many kinds of historical question, the version of realism with which we have ended up is in fact a natural concomitant. For we can pose two more conditions of adequacy on at least some (though certainly not all) historical research—conditions that are hard to make sense of without assuming what I have called a realist style of reasoning.

First, I think *we do an injustice to what drives much of contemporary writing in history if we underplay the degree to which it is driven by an interest in capturing what Tim Mason termed "the horizon of possibility" of the past*.[42] As Joyce Appleby put it (in her memorable response to David Harlan):

> If . . . we cannot fathom the original meaning of the texts offering us a window on other human experience, we will remain imprisoned in the present. Small wonder that historians draw upon their practice of reconstructing the past in order to resist this verdict.[43]

Now you do not have to think that we will be guaranteed success in the project of "fathoming the original meaning" in order for me to make my point. Rather, *if* you have a commitment to seeking such meaning, I do not see how to do so without taking a realist stance. The reason is that to take the horizon of possibility of the past seriously means that you allow that it shaped people's lives. That "how people saw the world, how they understood their experiences, how they saw their alternatives" are all elements of the historian's model that get carried over from what we take for granted from our world and elements of our world to which we already have realist commitments. We do not approach these features of our world with a theoretical attitude "as if"—we do not treat them as complex theoretical constructs appropriate to an instrumental or calculational model for understanding our world. Instead, we treat them as uncomplicated pieces of the furniture of our world.

Now of course you can choose to be a skeptic about this furniture—including the whole notion of original meaning. Some historians (most notably Joan Scott) have recently questioned the value of taking notions like experience as central to the historian's project. Scott proposes a wide-ranging hypothesis about the socially constructed nature of the categories of experience. She may be right—although I doubt that social constructivism will turn out to have anything like the reach that she thinks. But even if she is right about that, *and* right in claiming that "The project of making experience visible *precludes* analysis of the workings of this system [that gives rise to it] and its historicity . . . ,"[44] I don't see how Scott can pursue *her* project without being a realist as well. For, if your interest is in how experience (or anything else for that matter) is constituted out of features of the cultural and social system and in how these features are a focus of struggle between competing interests in that system, you have to take those features seriously as features in the world. There is a causal claim lurking here. For to speak of "construction" is already to tap into the vernacular of our causal talk. That is to say, there is a second condition of adequacy on historical research to be had: *to take the social construction of our lives seriously means taking it seriously as a causal feature of the world. And in this causal theorizing we display our realist commitments.* For here, too, as in the case of experience, we are all too familiar with how such notions operate in our own world, and we import such notions into historical models not as instrumental features of it but as elements of the world that are deeply implicated in shaping its contours—elements that are accorded a realist interpretation as part and parcel of our attributing the effects that we are interested in attributing to them.

Let me summarize the position I aim to defend: I say that, if we look to historical practice, it is hard to make sense of a lot of it without taking both the notion of historical perspective and causation seriously under a realist interpretation of the kind I have offered.[45] If my defense succeeds, realism will be shown to be compatible with positions that are often taken as its antagonists—the interpretative nature of history and the struggle against the hegemony of only telling one "Great Story."[46]

In the next three chapters I want to elaborate on these claims by illustrating how they are exhibited in a variety of different settings. The settings are framed by the methodological and theoretical views of others, and the methodology and theory is by and large *hostile* to the point of view I have been defending. Nonetheless, I am going to argue that, if we attend to the dictates of historical practice, the version of realism I have defended wins out.

Let us begin with the facts again. As I indicated in the Introduction, one notable feature of Novick's account is how few challenges to the view that leaves facts sacrosanct have come from within the historical profession, bracketing intellectual history.[47] But the same is far from true for other disciplines. Novick himself variously attributes the move against the status of facts in history to philosophers like Kuhn, Putnam and Foucault;[48] literary critics like Fish;[49] critics of the neutrality of the social sciences,[50] and the influence of Geertz.[51] Thus the indictment of a realist view of historical facts can be assayed from a variety of initial vantage points that are parasitic on a variety of other traditions. It will come as no surprise, given the foregoing, when I write that, of these, among the most important are views about the nature of facts that derive from considerations of language. I will examine versions of these views as they figure in writing about history in later chapters. But, first, I want to examine a position that argues for the indictment of facts for reasons internal and allegedly intrinsic to history itself, although here, too, as we will see, considerations of language quickly outpace considerations of history.

2

HISTORICAL FACTS

Can facts remain pristine, insulated from the anti-realist's scrutiny?

Even if there is a tradition about history that allows for the givenness of facts themselves, when it comes to *writing* history, how much good will this do us? Not much, if Hayden White is right in his claim that the narrative structure of historical writing renders *all* facts fictive because of the requirements of coherence.[1] But, the requirement of coherence notwithstanding, it does not follow from this alone that realism about the facts cannot be maintained. After all, *a priori*, there is no reason to think that the dictates of realism and considerations of coherence are incompatible.

Can we leave the status of facts undisturbed when we turn to consideration of writing history? In a very important book, Arthur Danto has argued that we cannot.[2] Danto is famous for an argument that aims to establish a formal incompatibility between historical knowledge of the past and the past itself. What makes Danto's argument a radical one is that it also aims to undermine the very notion of the givenness of the past itself—the core assumption that motivates the "Rankean" conception of history.

The picture of capturing the past "as it really was" is one that implies comprehensiveness as an ideal. There is a powerful temptation to think of the problem of historical realism as merely a function of distance. As such, a chronicler (who is by definition contemporaneous with the chronicled events) ought to be close enough to avoid such problems. However, the chronicler has another problem. Comprehensiveness is impossible in practice—it is not possible to document *everything* as it was. Practically speaking, then, chronicling the facts cannot realize the "Rankean" conception. A chronicle cannot be a complete rendering of the past as it really was. But what if we idealize the chronicle and allow that there could be a chronicler who chronicles *everything*? Our own inability to chronicle everything may simply be an artifact of our own limitations. An idealized chronicler may be thought of as simply lacking such limitations. As such, he or she could be thought of as representing the realist ideal.

Now one can argue against this view by holding that history has interests other than the chronicle, but Danto's line of attack on the relationship between chronicle and history takes a different tack, by taking aim at the coherence of the idealized conception of the chronicle itself and with it the conception of the historical that underlies it. Two related themes comprise the core of Danto's attack on the idealized conception of the chronicle and, in different ways, they both revolve around the centrality he gives to the notion of narrative. (Formally speaking, for Danto, narratives are composed of narrative sentences, and narrative sentences have special characteristics. But these characteristics need not concern us right away.) At the most general level:

> Any kind of narrative, assuming there were kinds of narrative, would require and presuppose criteria of relevance in accordance with which things would be included and excluded. This means I think that the maximally detailed account, the ideal of history-as-actuality, would not be narrative.[3]

Hence, *if* all history is narrative in nature, to the extent that the ideal is realized, it will not be history. Call this "the first incompatibility thesis" between narrative and the ideal of chronicle. If it holds true, whether or not the ideal of chronicle can capture "everything," it is irrelevant as an ideal for history. However, notice that for the first incompatibility thesis to hold, it needs to be shown that (a) all history is narrative and (b) narrative is of necessity selective.

The first incompatibility thesis notwithstanding, Danto also has in mind a much more complex and subtle argument that also holds that the ideal of chronicle and narrative must of necessity be incompatible. Call this "the second incompatibility thesis."[4]

The first incompatibility thesis exploits the comprehensiveness of the ideal conception to argue that it is incompatible with the selectivity of history. On the other hand, the second incompatibility thesis exploits the same feature of the ideal conception in the opposite direction, by showing it to be incompatible with the notion of what counts as comprehensive chronicle. It does so by taking the claim to comprehensiveness at face value and by then providing a radical interpretation of what falls under its domain:

> I [Danto] shall say now, that you cannot give a complete description of any event which does not use narratives. Completely to describe an event is to locate it in all of the right stories, and this we cannot do. We cannot because we are temporally provincial with regard to the future.[5]

So, while the first incompatibility thesis trades on the selectivity of narratives taken one at a time and counterpoises it to the comprehensiveness of ideal chronicle, the second incompatibility thesis runs in the opposite direction, counterpoising the incompleteness of ideal chronicle with the comprehensiveness of all narratives taken together. A complete description of the past on this view demands attention to all of the narratives in which the past may figure, and since we do not know the future, we cannot know all of these narratives. Hence, if completeness of the past does include all narratives that can be written about the past, then the ideal chronicler's task would seem to be unrealizable in principle—except in so far as he or she has foresight as well. But foresight is not a condition required of a chronicler. Nor is it part of our conception of the chronicler that we appealed to as a way of capturing an ideal for history that could capture the past as it "actually was." The second incompatibility thesis, then, asserts a condition that cannot be satisfied according to this argument.

Like the first incompatibility thesis, the second incompatibility thesis makes an assumption: that (c) the past is not fixed and unchanging.

Assumptions (a) through (c) need to be examined with some care, for they consist in the price that must be paid for buying the two theses.

ASSUMPTIONS (A) THROUGH (C) ELABORATED

It may seem so patently obvious that (a) all history is narrative as not to be in need of argument, and indeed Danto takes his "philosophy of history [to be] conceived of as a theory of narrative representation."[6] In the course of the argument, Danto poses what he takes to be a conditional not in need of argument; namely, "If we consider it to be the aim of historians to write narratives . . . ."[7] This is not to say that Danto thinks that history is *exclusively* composed of narrative. He allows that an historian might make a true statement about the past and that such a statement is not to be understood as consisting of a narrative.[8] And he allows that "There are descriptions of the past other than narrative ones."[9] Still, for Danto, implicitly these are the exception when it comes to history.

What of (b)? If (nearly) all history is narrative, then all history is selective. For Danto, the notion of narrative as anything but selective makes no sense. By its very nature, narrative imposes order by imposing a standard of relevance or selectivity according to which only certain events are selected.[10] Thus Danto takes it to be self-evident that: "When I say, then: 'tell me the whole story, and leave nothing out' I must be (and am) understood to mean: leave out nothing significant . . . ."[11] For to do otherwise would be to aspire to use narrative to obtain exhaustive knowledge—in some interest-free sense. But, "What, for instance, would it mean to have perfect knowledge of the Empire State Building?"[12] It would mean listing every fact and feature about it. But these are inexhaustible in number.

Assumption (c) is a much more complicated claim. The "Rankean" conception of history implicitly assumes that there is a past that we can aim to "get right" in our historical accounts. That is not to say that we have to embrace the metaphysical claim that (in some sense) the past exists. But the conception is informed by a view of the past as at least fixed. Common sense supports the same view: what is done is done. One way to undermine the realist view of the historical is to attack this idea. For if this idea falls, what falls with it is any notion of the completeness of the past along with any special deference to a chronicler of the past—even an ideal one. Danto characterizes the implicit model that the realist appeals to as follows:

> Let the Past be considered a great sort of container, a bin in which are located, in order of their occurrence, all of the events which have ever happened. It is a container which grows moment by moment longer in the forward direction, and moment by moment fuller as layer upon layer of events enter its fluid, accommodating maw. The forward lengthening of the Past is irrepressible, and regular; and once within the container, a given event *E* and the growing edge of the Past recede away from one another at a rate which is the rate at which Time flows. *E* gets buried deeper and deeper in the Past as layer after layer of other events pile up. But this constantly increasing recession away from the Present is the *only* change *E* is ever to suffer: apart from this it is utterly impervious to modification. . . . *E* and its contemporaries constitute an exclusive class, in the sense that no *further* event will ever join them, so to speak, a new contemporary. So the Past is not to change either through any modification of *E* apart from its momently increasing pastness, or though the addition of some other event contemporary with *E* which *E* lacked as a contemporary upon its entry into pasthood.[13]

I have quoted this colorful passage at length because it forms a crucial backdrop for understanding Danto's affirmative views about events.

What is wrong with thinking about the past as fixed and unchanging? Danto's surprising answer runs as follows:

> there is a sense in which we may speak of the Past as changing; that sense in which an event at *t–1* acquires new properties not because we (or anything) causally operate on that event . . . but because the event at *t–1* comes to stand in different relationships to events that occur later. But this in effect means that the *description* of *E-at-t–1* may become richer over time without the event exhibiting any sort of instability . . . .[14]

The idea, then, is this: the descriptions of an event that are true of it are dependent on the relationships that obtain between that event and other events. But some of those other events have yet to occur at the time of the event under description. That it is true of an event *E* at *t–1* that it anticipated an event *E* at *t–2* is true of

E at *t–1* but not available at *t–1*.[15] That the description that the author of the *Principia* was born in 1642 is true but not available until 1687.[16] Thus a complete description of events is not available to contemporaries of those events. Knowledge is available to the historian that is not available in principle to even an all-knowing contemporary (except in so far as that contemporary has foreknowledge).

Let us also note a more general feature of narrative: on Danto's construal, narrative sentences are sentences referring to distinct, time-separated events—they are, as it were, forward-directed.[17] But, strictly speaking, that renders *all* narrative sentences as closed to a chronicler. At the time of the occurrence of the first of the events that end up in narrative sentences, the chronicler cannot write down, as part of the complete description of the first event, its relationship to all of the second events.

Things are even worse for the chronicler. For if the aim is a complete description of all of the properties of an event *E* at time *t–1*, then the appearance of *E* in a narrative sentence itself is one such description. But there is nothing to rule out narrative sentences that include *E* being constructed at a later date as a function of historical interests. So a complete description of the past will be an artifact of our presentist interests. Hence Danto's clever twist that: "Completely to describe an event is to locate it in all of the right stories . . . "[18] where rightness is to be understood as an artifact of our interests. Completeness is then a function of something that *we* do.[19]

ASSUMPTIONS (A) THROUGH (C) SCRUTINIZED

If these arguments are correct, it would seem that there can be no complete realist account of the facts of the past, at least as understood in the straightforward way (which treats the past as fixed and independent of later events). But are these arguments correct?

The arguments, as they stand, place narrative at center stage, and so it might seem as if the argument could be undercut by undermining the claims made for narrative. Here I have in mind cross-cultural and cross-temporal historical studies of the kind exemplified by Natalie Zemon Davis's well-known study of the phenomenon of male–female festive role inversions that spans over three hundred years and at least six cultures.[20]

Instances like these would seem to undermine the claim of the centrality of narrative for history. I do not claim that these works have no narrative; rather, that narrative is not important to the structuring of these kinds of works.[21] In some cases this is true in part because their primary focus does not involve change. In other cases, and here I have in mind the Annales School, it is because of the long-term nature of the change. Perhaps one might argue that such cases are not in fact cases of history—but that is an implausible strategy given their prominence in the recent history of the profession. Instead, perhaps one might argue that they are in fact cases of narrative—even if not narrative as traditionally conceived. Here I have in mind a suggestion by Alan Megill that we allow for models of something close to non-temporal narrative. His idea is that narrative is composed of a number of different elements, of which temporal order is only one. Radically de-emphasize it and a work can still have many of the other elements of narrative: action, happening, character and setting.[22] Still, such a conception of narrative will lack the elements of which Danto makes use and render it compatible with a "Rankean" conception of facts.

Suppose that I am right and that it is correct to say that Danto has overemphasized the centrality of narrative—at least in the sense of its being organized around temporal order. Is this fatal to his argument against facts? It clearly is for the first incompatibility thesis. But what about the second? Here it will be for Danto, given the significance of the role that he gives narrative in generating both the revisable features of events and their forward-directed nature. But I want to claim that these features of events, if they hold true, in fact hold with or without narrative. Instead, they depend on a much more general claim.

Ignore the issue of narrative altogether and we are still left with the issue of whether a complete account of an event *E* occurring at time *t* can be given at time *t*. At base, Danto's reason for saying it cannot does not depend on a feature that is *unique* to narrative.[23] In narrative sentences, events at one time become related to events at later times (both directly by the forward-directed nature of narrative sentences and indirectly by the additions of new narrative sentences that further implicate the events) and that these relations obtain becomes the subject matter for true propositions that can be asserted about these events. But the truth of these propositions is not available to chroniclers.

But all of this can hold true without narrative playing any role at all. For the argument against the possibility of the ideal chronicler depends only on the fact that descriptive statements referring to events contemporary to the chronicler can be made at later dates. Such statements may involve the relationship between those events and later events. But even simple reference is enough to produce the problem. Thus it is true of the event of Martin Bunzl's birth that it preceded the writing of these words by forty-seven years, eight months and eight days. But this is not a truth that was available until this day of the writing of these words. So that now simply referring to the event of my own birth becomes more grist for the mill of the set of true (albeit narratively insignificant) propositions about the event of my birth: namely, that it was referred to on 11 March 1996.

Danto's claim comes down to this: the set of true propositions about an event is always expanding. And you don't need narrative sentences to make this case. You don't need to implicate events in new "stories" either. Just mentioning them is enough! *But does that mean that the event itself is changing, thereby undermining the idea of the fixedness of the past?* One way to avoid any claim that it does is to render such claims as purely epistemic. If it is true of the event of my birth that it has the properties indicated in the last paragraph now, then it was always true. The set of true propositions is not expanding. The set of propositions accessible to chroniclers is what is expanding. But even though Danto sometimes tips his hat in this epistemic direction,[24] in the main he defends the full-bodied ontological claim that a change in the set of descriptions true of the past generates the "retroactive re-alignment of the Past."[25] But what does this come down to? The relationship between the events in the past and their descriptions that Danto wants to forge is as follows. If we really take the notion of a full description seriously:

> We can imagine a description which really is a full description, which tells everything and is perfectly isomorphic with an event. Such a description will be definitive. . . . It now hardly matters whether we talk about the Past or its full description.[26]

But, given the openness of the future, there is no sense in which a "full" description is ever to be had in hand. What is really at issue is how we should think of the status of historical events and the stock of descriptions of them that is ever growing. Under such circumstances, do the events themselves change? Why not take events to be independent of their descriptions so that changing the latter will not change the former? Then the past will be fixed and unchanging, irrespective of the reorganizations of descriptions about it that take place over time. But then in what sense can there be "retroactive re-alignment of the past" unless we allow events to be description-dependent?

EVENTS AND FACTS

For better or worse an enormous literature lurks behind these two alternatives about how to construe the relationship between events and their descriptions.[27]

The classic polarity can be posed as the issue of how events should be individuated. Consider and contrast these two sets of examples:

(1)
The event of my drying my hair
and
the event of my drying my hair vigorously
and
the event of my drying my hair vigorously with a towel.

And:

(2)
The event of my swinging the hammer
and
the event of my breaking the vase
and
the event of my scaring the cat.

Are these cases of singular events that are differently described or multiple events for which there are differing descriptions?

Cases like (1) support the first interpretation. Here there is just one event about which we can provide increasingly detailed descriptions. The event of my drying my hair was the event of my drying it with a towel and so on. That is to say, the event can be picked out under a variety of descriptions.[28]

Case (2) looks like case (1)—my swinging the hammer *was* the event of my breaking the vase and that *was* the event of my scaring the cat. True, the scaring of the cat was not temporally simultaneous with my swinging the hammer; it probably came a little after. But notwithstanding this temporal order, by swinging the hammer I broke the vase and by doing that I scared the cat. Why can't we think of *these* as one? (My breaking the vase was my scaring of the cat.) The problem here is "by." It designates a train of actions from my swinging the hammer to my scaring the cat. If (2) can be rendered as a train of actions, then how can it also designate *one* event? The intuition that it ought not to, if there is a train of actions, is one of the motivations for treating events as description-sensitive (or more enlighteningly as property-sensitive). And on this view (2) can be thought of as a set of different events: instead of one event differently described, here we have different events that are picked out by different descriptions.[29]

Other cases of events that can be described in differing terms cut in different directions between these two alternative conceptions. When Socrates died, Xanthippe became a widow. She was widowed by his dying. But was the event of his death the event of her widowhood, or did the event of his death (perhaps) cause the event of her widowhood? The first alternative views the situation as one event differently described. The second views the situation as two events—unless we countenance self-causation. I doubt that there is a straightforward answer here that will mediate between the two alternative conceptions of events that are at play. In this example, much of what is at issue hinges on how we decide to conceive of widowhood. If you think of widowhood as something that happened to Xanthippe and death as something that happened to Socrates, then it seems counter-intuitive to treat them as identical, especially if Socrates and Xanthippe were not physically contiguous. On the other hand, if widowhood is conceived more as a state of affairs, intuitions push in the other direction. Think of widowhood less as something that happened to Xanthippe but more as a state of affairs that was rendered true. Then the state of affairs that rendered it true that Socrates was dead is also the state of affairs that rendered it true that Xanthippe was a widow. "Rendered" here is not causal. Rather it is the state of affairs that obtains in virtue of which the statements "Socrates was dead" and "Xanthippe was widowed" are true. *And it is the identical state of affairs in the world in virtue of which they are both true.*

Note that this is not to say that Socrates' death and Xanthippe's widowhood could not have been rendered differently, and, more importantly, they are necessarily rendered true by the same state of affairs. For if Socrates had not been married to Xanthippe at the time of his death then the event of his death would not have been the event of her widowhood. In this last thought there is an outline of a counterfactual approach that could be used to show that Socrates' death and Xanthippe's widowhood ought to be thought of as two different events even if they happen to go hand in hand. But I don't want to consider the advantages and disadvantages of this approach here.[30]

With or without modal considerations, the case of widowhood illustrates another issue that is lurking in the background. We began by posing the discussion of Socrates' death and Xanthippe's widowhood as *events*. But then we slid into talk of *states of affairs*. With a suitably relaxed attitude toward the notion of an event, states of affairs can be thought of as a species of event. They are just kinds of events that lack the usual temporal and spatial compactness that we normally think of in connection with regular events. The event of the bridge's collapse is specific to a particular time. So too with the event of Xanthippe's *becoming* a widow. Contrast that to her widowhood. That state of affairs is an event too—just one that is smeared over an extended period of time.

But are we really talking about a smeared event here, or a fact? I take the difference between the (smeared) event of Xanthippe's widowhood and the fact of her widowhood to be this: the event encompasses all sorts of properties—it was (perhaps) sad, lonely, unhappy, relieving, physically dislocating, financially impoverishing and so on. The fact of her widowhood was just that. It simply refers to one of the properties of the event. But I said earlier that there is a state of affairs that renders it true that Xanthippe was a widow and what renders it true that Xanthippe was a widow is just one property or feature of Xanthippe's condition—a fact, not an event. So perhaps my use of states of affairs as assimilable to events was too hasty. "States of affairs" looks like it can be applied to either facts or events.

Once we recognize this ambiguity it becomes possible to

generate a variety of combinatorial alternatives of conceptions of facts and events. As such we need to face the question about whether what is at issue here in the discussion of the past is an argument about facts or events. We have seen that we can treat events as impervious to redescription or not and, at least in principle (even if it is philosophically unpopular to do so), we can treat facts likewise. Then, if you want a fully revisable past, simply pick the appropriately revisable account of both facts and events. Likewise, to ensure complete conservatism with respect to the past, pick the other conception of both facts and events. What about the two middle roads?

I take Danto to be implicitly choosing one of them when he writes:

> there is a sense in which we may speak of the Past as changing; that sense in which an event at *t–1* acquires new properties not because we (or anything) causally operate on that event . . . but because the event at *t–1* comes to stand in different relationships to events that occur later. But this in effect means that the *description* of *E-at-t–1* may become richer over time without the event exhibiting any sort of instability . . . .[31]

Here we have what amounts to an account of events that are impervious to change by way of redescription combined with a revisionary account of facts. The events of the past remain the same but the facts true of them change.

To take the middle road is to take the permanence of events seriously.[32] Realism can obtain at the level of events, then, without extending to the domain of facts. The facts of the past can change. So far so good. But why not view the changing status of facts in purely epistemological terms? Then we can give up any mysterious notion of the past changing. Why buy into the trouble of saying that it is the past that is changing rather than our knowledge of it? Think of facts about the past as matching the set of true propositions about it. If we think of them as atemporal, then the set of actual facts is fixed, it is just our knowledge of them that changes over time. But it seems to me that Danto is right in insisting on the *essentially* temporal nature of history. Once this temporal dimension is admitted, *relative to it*, some things at time t are "not yet facts" and the epistemological move is blocked for all intents and purposes.

But can we think of the "Rankean" ideal as realizable at the level of events even if not at the level of facts? The problem with this conception of history is that we have to make sense of what it is to write a history of events rather than facts—even smeared-out events. The problem obtains even for the most straightforward and mundane instances—the favorite examples chosen by philosophers. Consider the event of Caesar crossing the Rubicon and consider writing the history of the event. The problem is this: there is no medium for such a history except by way of facts. Try to write the history of the event without facts and how will you write about it? You can refer to the event. But beyond that you will immediately stray into facts. The reason is not hard to see: (if successful) your writing will be in the form of descriptive statements true of the event, and doing that is precisely what will pick out facts about the event.

The upshot of this argument is that the existence of a middle road that is realist about events but not about facts is an illusory one for history because history is (at least directly) not about events but only about facts. We may want to write about events, but all we can write about are facts. Events are one step removed. As such it is the character of facts, not facts and events, that will determine the character of historical writing. Hence, the "stability" of events notwithstanding, from the point of view of writing history we *can* treat the past and description of it as equivalent to the extent that description of the past becomes ever-enriched over time, the past is *not* fixed and unchanging. Hence, there is no "Rankean" ideal, at least if it is understood in the sense of contemporaneous chronicle.[33]

Be that as it may, there is a temptation to think we can go further in distinguishing between facts that do not change and those that do. For example, the event of World War II might be thought to be composed of some facts that constitute a class whose membership is fixed—at least at the war's end. On the other hand, the class of facts about World War II is continually expanding in virtue of relational properties that World War II bears to later events and facts, for example, to the fact that Martin Bunzl was born three years after the war's conclusion. That fact, however (when the circumstances that made it came to be), made no change in the war itself. Can we draw a line between the "real" facts that make up World War II and mere relational facts? As a practical matter, distance may give rise to comparative stability when it comes to facts. Still, I see no way to make a distinction between "real" facts and relational facts work at a principled level when it comes to history. The obvious and, to my mind, only relevant way to distinguish between kinds of facts about an event is to do so causally. World War II is related to my birth but impervious to it because (barring backward causation) my birth could have no effect on World War II. And while it is true that the birth of the author of the *Principia* was not such a birth until Newton authored the *Principia*, that fact is causally impotent when it comes to Newton's birth.[34] But there is no reason to think that our historical interests are in any sense restricted to just the causally efficacious features of the world. Hence we will be hard pressed to distinguish between different kinds of facts by way of reliance on causal notions.

SELECTIVITY AND THE PAST

Still, how much bite does the position we have ended up with have against a realist conception of the past? In particular, where does it leave the status of facts that embrace the perspectives of historical actors?

We have been treating the "Rankean" conception of the past as, among other things, embracing two related principles: the fixity of the past and the completeness of the account of the past. Let us concede that neither of these can be held as ideals on the argument considered so far. The facts are not fixed and accounts of the facts must of necessity be selective. But what happens if we restrict our interest to providing accounts of the interpretations of historical actors? Closed to the future, historical actors are of necessity bound to be selective with regard to the facts by their own "horizon of possibility." But what about our pursuit of their accounts? If our interest is in capturing their world, then, *relative to that interest*, why can't we aim to achieve completeness? For, relative to that interest, the facts *are fixed*. They are fixed in virtue of the historical actors' own selectivity.

The problem of how to deal with the issue of selectivity may only seem to arise if we aspire to an ideal of completeness under circumstances which we have seen to be impossible. Nonetheless, Danto's argument still allows us a different sense in which to express the natural prejudice that, in matters of history, more is always better. As events recede, they become increasingly implicated in their relationship to events that follow and we have more facts available to us about the past. Why should this be anything but a virtue? If our interests are limited, why not simply be selective about our knowledge? Yet doing so requires great discipline if our interests are in the phenomenology of historical actors. For the difficulty of looking back is that it distorts the limited perspective of those trying to look forward and hence can easily distort our understanding of what it was like for them.

German history offers a telling example.[35]

On 30 January 1933, Hitler came to power. Between that date and Kristallnacht, some five and half years later, there was no wholesale emigration of Jews from Germany. In fact only one third left, and of those some had actually returned by the time of Kristallnacht.[36] Indeed, the period was one in which new Jewish cultural institutions where established, while one leading member of the Jewish community called for the yellow star to be worn with pride.

How can we understand these behaviors? Certainly we cannot erase the ensuing history from our minds. Yet, if our interest is to try to reconstruct the significance that contemporaries placed on this five-year period, our aim must be to write a history that excludes facts about the events of the period that were not available to them at the time.

Recent scholarship about the period offers an account of how Hitler came to adopt the goals of the militant wing while imposing a more rational method as a way of achieving these goals. It offers an account of how internal party politics and the interest in placating the militants led to this change.[37] But, more important for our purposes, it is also a case which illustrates the way in which looking back to a period can confuse the picture by undermining the role of chance and accident in the etiology of events. And with the role of chance and accident undermined, we lose sight of how limited the horizon of possibility was for the actors involved.

On the view under discussion, Nazi policy came to be set as a result of a fight between two wings of the party: one wing was broadly anti-capitalist, anti-modernist, anti-establishment, *and* had as its ideological core anti-Semitism. This militant wing was largely based in the SA (Sturmabteilung). In contrast, the other wing had much broader interests, including foreign policy, preserving support for business elites, conciliating the press, maintaining the German import–export trade which involved keeping Jews in the economy, and not alienating the middle class. On this view of Nazi anti-Semitic policy during the period 1933–1938, nothing was foreordained. Instead, policy was determined year to year as an *ad hoc* response to changing conditions and the balance of power inside the party.

The predominant effect of this state of affairs was chaos. In 1933 some contemporaries thought the Hitler government would last no longer than the string of governments that had fallen in the previous three years. Reading contemporary newspapers and journals one could hardly conclude that there was anything like a consensus that Hitler's Reich would last for twelve years.[38]

In the first months of the regime, as institutions of civil society were systematically destroyed, those physically removed to detention and work camps were primarily prominent socialists and communists.[39] Because Jews had been prominent in these parties, it came as no surprise that Jews constituted a high proportion of those interned during this period. Perhaps what has been lost over time is the perception that it was not in virtue of their Jewishness that these people were interned.

Even the boycott of Jewish stores on 1 April 1933 did not give the message that the militant wing of the Nazi party would eventually set the policy toward Jews. The militant wing intended the boycott to last indefinitely. However, Hitler's early policy of restraining his militant supporters resulted in a one-day boycott, and indeed the elimination of members of the SA followed in June of the next year with the Night of Long Knives. These killings notwithstanding, the militant wing was not eliminated. But it was not until Kristallnacht in 1938 that it became clear that the aims of the militants would in fact be achieved, even if it turned out to be not by way of the militants' methods.[40]

THE PROBLEM OF REFERENCE

I have been arguing for taking some historical facts as fixed and complete—facts about the interpretations of historical actors. These are not all the facts that form the content of history. Nor are they all the facts about the interpretations of historical actors in which we may have an interest. If we try to pursue generalizations about the status of all of the facts of history, the issues of fixity and completeness rear their heads. But I have been suggesting that this is not the case, at least for one slice of our historical interests.[41]

Yet, notwithstanding this argument, we are left with a different problem that even threatens the restricted domain of facts we have been considering. Assume that there are (fixed) facts about the interpretations of historical actors. Still, are they available to us?

In the next chapter I want to examine a set of influential arguments that such facts, and in general all facts about the past, are not so available to us. The reason is not because of the nature of facts or because of considerations of epistemology, but rather because of the nature of how we try to refer to those facts in the construction of history.

3

THE CONSTRUCTION OF HISTORY

If there is a major lacuna in *That Noble Dream*,[1] it is that Novick's history misses the turn towards textuality among some American social historians. The fault is not Novick's, but just unfortunate timing. His book was completed and its publication coincided with the appearance of Joan Scott's much discussed volume, *Gender and the Politics of History*.[2]

Scott's theoretical essays in that volume are of special importance because, more than perhaps any other published work in the discipline, they provided a manifesto for the incorporation of post-structuralist philosophy into the writing of social and cultural history.[3]

In this chapter and the next one, I want to examine this manifesto and others related to it. For, taken together, they constitute a call for an approach to history that radically undermines any claim to the fixedness of facts. They do so, not because of any special arguments about the nature of facts themselves, or of history, but in the end because of a view about language and its use in the writing of history.

These manifestos exist sandwiched between two formidable bodies of work. On the one hand, they are written with enormous and self-conscious attention to the work of both Derrida and Foucault. At the same time, they coexist with a body of historical practice driven by quite stable and long-standing concerns to write the social and cultural history of those whose history has not been written.

As such, evaluating these manifestos is complicated by the task of deciding whether to judge them against the philosophies that catalyzed them or against the backdrop of the practice that is said to instantiate them. I will try to do both, and I will argue that, against these standards, the results are disappointing. But, in the end, I will try to argue that examining just why these results are disappointing can yield a different strategy for trying to evaluate the relationship between theory and practice, and, with it, the status of historical facts as well. I begin with Scott.

SCOTT

Scott's methodological views are driven by a substantive thesis about categories in general and gender in particular. Her interest is in:

> historicizing gender by pointing to the variable and contradictory meanings attributed to sexual difference, to the political processes by which those meanings are developed and contested, to the instability and malleability of categories of "women" and "men," and to the ways those categories are articulated in terms of one another, although not consistently or in the same way every time.[4]

Here, gender is not to be understood as an historically "stable" concept about which we can generalize, and treating it as if it were stable ends up as a self-defeating approach to writing the history of women. Simply uncovering new facts about women, argues Scott, does nothing to overthrow the marginalized treatment of women as historical subjects.[5] To do more means changing the very categories that are used to organize history, as well as making them objects of historical study.[6] "Gender" for Scott means "knowledge about sexual difference," whereas "knowledge" means "the understanding produced by cultures and societies . . . ."[7] As such, gender and, with it, knowledge are matters of social organization, and this social organization is the object of study, including the role of history itself as a discipline that is implicated in the process of social organization.[8]

Studying such social organization, Scott enthusiastically embraces elements of post-structuralist method to capture the role of rhetoric and discourse in which "meanings [of concepts] are not fixed in a culture's lexicon but are rather dynamic, always potentially in flux."[9] We can see in Scott three elements drawn from this post-structuralist approach. First, an appreciation of the conflictful nature by which competing interests in a culture are in contest about the meanings of concepts. Second, a commitment to studying the way in which the definition of concepts is contrastive. And third, an anti-universalist stance toward the meaning of concepts.

A CRUCIAL CLAIM

As things stand, there is no reason why gender *could* not turn out to be a stable concept about which we might generalize. Still, Scott thinks that the pursuit of our interests in gender demands a new "epistemology"[10] in which "knowledge is not absolute or true, but always relative",[11] and it is this conception of epistemology that is the key to understanding her theoretical views, including her claim "that universal explanation is not, *and never has been possible*."[12] For Scott sees that this epistemological view can be applied to the writing of history itself, and she is willing to embrace what she sees as the consequences of this move:

> It also undermines the historian's ability to claim neutral mastery or to present any particular story as if it were complete, universal and objectively determined. Instead, if one grants that meanings are constructed through exclusions, one must acknowledge and take responsibility for the exclusions involved in one's own project. Such a reflexive, self-critical approach makes apparent the particularistic status of any historical knowledge and the historian's active role as a producer of knowledge.[13]

Of course it is just here that the clash with realism about historical facts seems to occur with full force. This is not necessarily to deny that there is a past or that there is no way that things were in the past; rather, it is just that that is not the subject of historical discourse. And one way of achieving this separation is by taking a resolutely linguistic turn and emphasizing the textual features of historical writing instead of its referential features.

Reading Scott, there is a temptation to see her as advocating this linguistic turn in some of her theoretical writings when, for example, she writes that:

> history's representations of the past help construct gender for the present. Analyzing how that happens requires attention to the assumptions, practices, and rhetoric of the discipline, to things either so taken for granted or so

> outside customary practice that they are not usually a focus of historians' attention. These include the notions that history can faithfully document lived reality, and that categories like man and woman are transparent. They extend to examinations of the rhetorical practices of historians, the construction of historical texts, and the politics—that is power relationships—constituted by the discipline.[14]

I have quoted Scott at length here because of the counterpoint that is posed between what is "usually taken for granted" [!] as part of the historian's agenda and the alternative inspired by post-structuralist considerations. Some commentators have taken this alternative as signaling a complete turn inward of history. Thus, in a review of *Gender and the Politics of History*, Linda Gordon characterizes Scott as arguing that "language is and must be the only subject of history . . . ."[15] This is surely a caricature, and it is one that Scott is quick to reject.[16] But nonetheless it points to the issue of to what degree, in making language central, the aim is to stop at language or reach beyond it. Certainly critics of the introduction of post-structuralist theory to social history have worried that, if not the aim, at least the result of its application will be to stop at and not reach beyond language.[17] In response, Scott accuses Gordon of "conflat[ing] signification (the way humans construct and express meaning) with 'language' which is, in her usage not mine, merely 'words'."[18]

Implicit in this response is the view that there *is*, then, a notion of meaning to be had—one whose construction can be chronicled.[19] The problem, however, is just how to reconcile such interests in practice with Scott's views about this very process of signification in the writing of history itself.

PRACTICE

In "On Language, Gender, and Working-Class History,"[20] Scott offers a critique of Gareth Stedman Jones's work on Chartism.[21] Scott admires Stedman Jones's aspiration to avoid seeing Chartism merely as a political movement based on a given social reality dictated by a theory of material political economy. Stedman Jones's work is in part a remarkable plea for attention to be paid to the way in which language plays a central role in, as Scott puts it, "providing an interpretive definition for experience within which action could be organized."[22] But whatever the promise of Stedman Jones's plea, Scott seems to me to be on the mark when she takes him to task for offering a political history in which, in the end, language is merely reflective of reality and not constitutive of it.[23] The contrast Scott aspires to offer is an account of the way in which the history of Chartism might have been written in such a way as to demonstrate the constitutive properties of its language, especially with respect to class and gender. For Scott the connection between gender and class in the construction of nineteenth-century "languages of class" turns out to be extraordinarily intimate. In fact, Scott asserts: "There is no choice between a focus on class or on gender; each is *necessarily* incomplete without the other. . . . The link between gender and class is *conceptual* . . . ."[24]

How should we assess what is going on here against the backdrop of Scott's stated theoretical commitments?

Of course, Scott's conclusion that there is a necessary and conceptual link between gender and class seems to be in contradiction to a central claim of her theoretical commitments. Words like "necessary" and "conceptual" are part and parcel of just the kind of claims philosophers usually talk about as being "universal," "fixed" and "unchanging." That is, as having just the properties that Scott wants to disavow for all talk, including historical talk. Should we take these claims at face value or not? To do so we have to conclude that there simply is a contradiction between Scott's claims in theory and in practice. There are other alternatives. One is that we are pressing too hard on the intended sense of "necessary" and "conceptual." Perhaps the intended sense does not carry any philosophical baggage and is rather used as a matter of rhetorical emphasis. Yet this seems like an interpretation of last resort in an essay that was authored, in the main, to make a methodological point about the role of gender in class analysis.

There is another, albeit messy, interpretative alternative. In order to understand the intended sense of "necessity" and "conceptual," perhaps we need to ask about the intended sense of "class" and "gender" when Scott asserts that: "There is no choice between a focus on class or on gender . . . ."[25]

If, speaking with the vulgar, we take these terms to be unrestricted in their domain of application, then the problem with the claim that they are necessarily and conceptually linked won't go away easily. I say "speaking with the vulgar" because Scott's intended sense of these terms is different. Class and gender, like all concepts, are historically constructed and, of course, her claim is that in the post-Enlightenment period, the construction of gender and class are interrelated. So perhaps we can think of them as *necessarily* interrelated in this localist sense.

So far so good. But there is still a problem: just because the concepts of gender and class have been interrelated, in what sense *must* they be so interrelated? Now one way to sidestep this question is to allow that they need not be interrelated, but in that case they would be different concepts. But sidestepping the question in this way has an important consequence: we cannot use a *general* notion of class or gender outside a specific historical context in which the terms have a "construction." To someone like Scott that may be a virtue of the account, not a vice. For, after all, she wants us to eschew the universal and embrace an account in which "meanings [of concepts] are not fixed in a culture's lexicon but are rather dynamic, always potentially in flux."[26] But there is a hidden price to pay for such an approach to meaning. It is a price that is quite consistent with what Scott embraces in theory. But not, I want to argue, in practice.

The situation is parallel to that found in Kuhn's approach to scientific change.[27] This is not accidental, for Kuhn, too, is a localist of sorts.[28] One of the consequences of Kuhn's way of thinking about scientific change as convulsive is that scientists in different periods are not talking about the same subjects. Mass is a different concept for Newton and Einstein, because Newton's theory includes no upper bound for how fast objects can travel, unlike Einstein's theory, and for Kuhn what counts as "mass" is determined by these theories. *Thus, to the extent that competing theories are different, so too is what they talk about.* To use a famous turn of phrase of Kuhn: "the proponents of competing paradigms practice their trades in different worlds."[29]

The philosophical problem with this sort of view is not necessarily the convulsive picture of paradigmatic scientific change *per se*.[30] Rather, the difficulty is that the theory of reference that grounds the claim that Newton and Einstein are referring to different things when they speak of mass is *too* powerful. Even if you want to allow that Newton and Einstein *are* talking of different things when they use the term "mass," unfortunately such a view of reference forces you to make the same claim about *any* theoretical difference. For, if the referent of a term is determined by the theory in which it is embedded, then the holistic nature of theories will make *any* difference in theory a difference in reference. But then two scientists with mildly different concepts of mass will be discussing different subjects when they use the term "mass" instead of having different concepts about the same subject. The result is that, inadvertently, all *scientific* change will be rendered convulsive because there will no longer be a concept of intra-paradigmatic continuity of reference. Yet, of course, it would be *prima facie* implausible to say of *all change* that:

> a new scientific truth does not triumph by convincing its opponents and making them see the light, but rather because its opponents eventually die, and a new generation grows up that is familiar with it.[31]

The difficulty of the view is brought out by thinking of what it is to move from an incorrect theory of, say, the electron, to a correct one. For such talk itself is ruled out by this view of reference as well—because there is no extra-paradigmatic view of the "electron" to be had. So there is nothing that can be successfully referred to as the object of interest of competing theories. In philosophy of science, for those who think that this is an indefensible position, the challenge has been to provide a basis for how reference can work to allow for just such an extra-paradigmatic account to work. That is, for a way to refer to things or concepts or whatever, independent of the theories that may be offered of those things.[32]

The parallel here for Scott's theory is obvious—there is no atemporal, ahistorical notion of terms like "gender" or "power" to be had. What these terms mean is in flux. But how does this translate into practice? To first appearances, very well. For the view provides a caution against making cross-temporal generalizations using a term as if the term were stable.[33]

Still, if we are true to such theoretical commitments of localist meaning, how should we interpret statements like part of Scott's concluding paragraph of her essay on Stedman Jones?

> It may be visionary to hope that a more sophisticated theory of discourse will also open the way for a needed

> reconsideration of the politics of contemporary labor historians. Many of these historians, writing from a position that supports the democratic and socialist goals of past labor movements, uncritically accept masculine conceptions of class and rule out feminist demands for attention to women and gender as so many bourgeois distractions to the cause.[34]

Surely, in this quotation, terms like "masculine conceptions of class" have an intended reference that cuts across the provincial loci within which particular constructions may take place, just as much as the reference of "electron" does across particular theories of it. The problem is the same as that faced by a theory of reference for science. But here I want simply to make a claim about historical practice, including Scott's practice: it has to assume something more than localism in its reference. Thus, however localist one's leanings may be, considerations of language will force one into something more.

Such referentially driven anti-historicism[35] comes in two different variants. One is a matter of the construction of history itself in which Scott's interest, and that of nearly every other practicing historian, is to generalize, and this is just what historicism rules out. (I say "nearly every other historian" because avoiding even minimal generalization is hard to do. With or without narrative, historical discourse is hard to generate if you are prohibited from talking about more than one instance of a particular *kind* of thing. Yet talk of "kinds of things" is generalized talk. Thus you cannot assert that "Gender is one of the recurrent references by which political power has been conceived, legitimated, and criticized . . . gender and power construct one another . . . "[36] without thereby committing yourself to the capacity of "gender" to pick out something that these recurrent references have in common in spite of their differing mutual interrelations with power.)[37]

But anti-historicism comes in another variant as well: just as full-blown Kuhnianism (if only inadvertently) quickly makes it impossible for scientists to understand each other, so the same follows for historians who embrace full-blown localism. For such localism will extend to historians' language itself. Yet the glaring fact is that scientists in practice do understand each other—even scientists working from different paradigms—and so do historians. And any successful theory of reference to be applied to either science or history needs to be consonant with this as a condition of adequacy.

The argument I have been giving for the contrast between Scott's theory and Scott's practice is a general one. It is one that I have suggested can be made of practically any instance of the practice of historians, given the need of such practice to refer beyond specific instances. But, for Scott, the contrast between theory and practice arises in another way as well, given her interest in writing the history of women. Scott approaches this history convinced of the need to write the history of the category of woman inspired by post-structuralist theory. Yet just how to make the latter work for the former is harder to do than it may at first seem. The central claim that I want to make is that putting post-structuralism to such work involves a substantial transformation of our understanding of post-structuralist theories themselves. As I will argue in the next chapter, in the case of Foucault, I think the result is a new interpretation; while in the case of Derrida, the result, though perhaps inspired by post-structuralism, is a distinct theory. But for now, let us look at the consequences for women's history itself.

THE CENTRAL DILEMMA

What is at stake can be best illustrated by considering an example such as Scott's essay, "Work Identities for Men and Women: The Politics of Work and Family in the Parisian Garment Trades in 1848."[38]

In 1848, both (male) tailors and (female) seamstresses were involved in campaigns about work. Pressed by increasing competition from ready-made clothes produced by home-based workers, the tailors' main interests revolved around the locus of work—especially, the maintenance of production in workshops rather than in the home.[39] In contrast, the seamstresses' primary interest was a matter of wages—whether paid for work in the home or in the workshop.[40]

For the tailors, shop and skilled production were to be contrasted to home and unskilled production,[41] and the attack on home-based production made use of images of both the family and women:

> The self-exploitation associated with homework was . . . seen as corrupting the order and emotional fabric of family life.[42]

And:

> The explicit objection to home-based production was that it violated separate types of male and female activity, and that it robbed family members of a control over their distinctive responsibilities.[43]

Although the seamstresses' demands were more general than those of the tailors, nonetheless,

> From one perspective, the appeals . . . were remarkably similar. . . . Both stressed the rights of producers and divisions of labor between the sexes, divisions that, although they associated women with home and family, did not make domesticity the antithesis of productive society.[44]

Still, there were important differences between the tailors and the seamstresses:

> While the feminist seamstresses offered wage-earning as proof of the fact that women qualified as producers (and thus as citizens), political tailors premised their collective identity on the possession of (historically transmitted) skills.[45]

By so defining collectivity, the tailors were able to preserve a hierarchy between themselves and the largely female wage-earners.

Nonetheless, these differences notwithstanding, both seamstresses and tailors used the rhetoric of the family to articulate their interests, contrasting them and the family to the effects of capitalism. The image of the family was thus put to use as a locus for the contrast between the cooperativeness of utopian socialism and the exploitative individualism of capitalism.

What made possible the use of this shared image, given the differences in the interests of the tailors and the seamstresses? Scott suggests the answer lies in the *ambiguity* with which the image of the family was raised:

> The family was projected as an abstract entity, a place of complete fulfillment, in opposition to the alienation of

> capitalist society. The resolution of conflict and competition was depicted in the heterosexual couple: a harmonious reconciliation of opposites masculine and feminine. Whether masculine/feminine was equal or hierarchically arranged within the unit was ambiguous in the vision; the ambiguity permitted the different interpretations developed by tailors and seamstresses, and also a certain inconsistency in usage.[46]

Here, then, we have a trenchantly drawn account of the variable and constructed use of the family in a particular historical context. It is an account that draws its power by making plausible just how the rhetoric of the family could be used to serve a collective interest, despite profound differences within that collectivity. But how does this instantiate the kind of theoretical interests that Scott defends?

What "Work Identities for Men and Women" illustrates so well is a "contextual reading that taps into politics from a *specific and popular perspective*."[47] Yet *that* is a very different interest from those articulated by Scott earlier. Here there is no "reflexive, self-critical approach [which] makes apparent . . . the historian's active role as a producer of knowledge."[48] And, in fact, the thrust of Scott's questions in "Work Identities" seems incompatible with the earlier warnings that we should not take for granted "the notions that history can faithfully document lived reality . . . ."[49] For if we can't take this for granted, then in what sense can we hope to realize the interest in realizing a "contextual reading that taps into politics from a specific and popular perspective?"[50]

How are we to reconcile the tension between the ambitions articulated by these two quotations?

Mary Poovey has faced this issue head on. In "Feminism and Deconstruction,"[51] Poovey's concern is in the relationship between deconstruction, as inspired by Derrida, and feminism. But the arguments apply with equal force to any post-structuralist program (including that of Foucault) that shares with the Derridean program the undermining of both truth and identity. I say this because, as Poovey puts it:

> To take deconstruction to its logical conclusion would be to argue that "woman" is *only* a social construct that has no basis in nature, that "woman," in other words, is a term

> whose definition depends upon the context in which it is being discussed and not upon some set of sexual organs or social experiences.[52]

This is, after all, precisely the kind of general post-structuralist manifesto that we have been examining. The problem for such a position, whether it is Derridean or not, as Poovey puts it, is that "This renders the experience women have of themselves and the meaning of their social relations problematic . . . ."[53]

Poovey's work is important because she takes the Derridean program of deconstruction seriously and yet she sees both the need and the problem in reconciling it with an experientially based history. She wants "to work out some way to think of both [the experience of] women and [the category of] 'woman'."[54]

For Poovey, the implication of deconstruction is that there is no category of "woman" except in relational terms to the category of "man" (and of course vice versa). This interdependence is one of words and, without any account of given categories *in nature*, it is language that forms the basis for those categories. But Derrida is a holist of sorts about language. The inter-definition of "woman" and "man" is part of a more general chain of signifiers, so that "None of the members of this linguistic chain has priority . . . ."[55] So truth as a correspondence relation between words and the world appears as an illusion.

For now, at least, I am not so much concerned with the details of this view as with its conclusion: "From the perspective of this project, a feminism that bases its epistemology and practice on women's experience is simply another deluded humanism . . . ."[56] In the Derridean mode, this view about epistemology arises because language allows no priority to be given to the category of "woman" as one organizing *experience* as opposed to one organizing theory. But we have seen that the road to the same epistemological conclusion can be reached without embracing this Derridean view of language. Instead, the more general conception of "knowledge . . . [as] not absolute or true, but always relative . . . "[57] will have the same effect with respect to the epistemological status of experience.

With this connection in mind, we can also generalize Poovey's argument in another direction as well—I think that her treatment of "the positive contributions deconstruction [makes] to . . . feminism . . . "[58] is also aptly applied to a more general conception of post-structuralist theory, at least in one respect, namely the possibility of "a genuinely historical practice—one that could analyze and deconstruct the specific articulations and institutionalizations of . . . categories, their interdependence, and the uneven process by which they have been deployed and altered."[59]

Still, whether this is viewed in more general post-structuralist terms or not, a parallel problem arises to that encountered in considering the category of "experience." Poovey realizes that this conception of a "genuinely historical practice" is not a part of a deconstructive program, it is something that moves beyond it. Yet can it do so in a way that is consistent with the assumptions of deconstruction?

Poovey's proposal is to call for an historicization of deconstruction itself. The results she suggests will be a transformation of the subject:

> Ultimately, my prediction is that feminists practicing deconstructive and other poststructuralist techniques from an explicitly political position will so completely rewrite deconstruction as to leave it behind for all intents and purposes, as part of the historicization of structuralism . . . .[60]

But, as far as I can see, this cannot stand merely as a prediction. For the problem Poovey poses pertains to work being done now: how *is* deconstruction (and more generally post-structuralism) in fact reconciled with an interest in both historically located *experience* and categories? Do these histories in fact synthesize historically located experience and categories with post-structuralism? And, if so, do they entail a transformation of post-structuralist theory?

4

FOUCAULT BY HISTORIANS

In "Foucault for Historians," Jeffrey Weeks provides an eminently succinct summary of the main elements of the Foucauldian program; while in *Sex, Politics and Society,* he attempts to develop an anti-essentialist approach to sex and sexuality in the context of British history.[1] If ever there was a topic well suited to a Foucauldian approach it is surely this topic, and one would expect "Foucault for Historians" to mesh neatly with *Sex, Politics and Society* if it does anywhere. So it is instructive to see why in fact this is *not* the case, and doing so turns out to be helpful as a first step toward a general argument about what it is to write history with an eye on the Foucauldian corpus.

To speak of "Foucault for Historians" is implicitly to assume that Foucault himself was not an historian, and certainly he was not one judged by conventional standards, nor was he by his own lights. Still, whether you want to call him an historian or not is not very interesting. Irrespective of that issue, the problem with recommending Foucault for historians is how to reconcile central elements of his program with the conventional interests of historians.

Foucault's program was explicitly presentist. But that in itself does not set it aside from the programs of many historians. Nor does Foucault's aim of studying the past as "curative" of the present set it aside from the work of many historians. But what does set it apart are the methodological consequences that Foucault drew from this interest: an explicit disavowal of the importance of causal determination in general and agency in particular. I say "Foucault drew" because the exclusion of these elements of historical interest does not follow from the fact that he held presentist interests *per se.* For our purposes here it will turn out not to matter why he held the view—although the most plausible grounding seems to me to lie in seeing it as an extension of his skepticism about the social sciences.

In "Foucault for Historians", Weeks acknowledges the acausality of the Foucauldian program, but he attempts nonetheless to steer around it while embracing both Foucault's anti-essentialism and the centrality of the study of discursive practices:

> the task of historical investigation is not to fish for the "real" history that glides silently under the surface, or rules behavior behind men's back, but to address itself to the surfaces, which *are* the "real" in the way we live social relations through the grid of meaning and language.[2]

That is to say:

> The task of historical investigation is the understanding of the processes of categorization which make phenomena like sex socially significant, and which produce the forms of knowledge which provide the focus for social regulation and control.[3]

Nor is acausality the only thing that Weeks wants to steer around in Foucault's work. Weeks also dislikes the lack of agency,[4] the lack of connections between discourses,[5] and the diffuseness of Foucault's concept of power.[6] Still, having said all that, Weeks recommends Foucault as "a valuable, if necessarily partial guide,"[7] and I think that he is correct in this assessment. The problem is how to make selective use of Foucault as a guide, and one of the ways to see this problem is in Weeks's own work on sexuality.

In *Sex, Politics and Society*, an overview of the "regulation of sexuality" from 1800 to the present in Britain, Weeks's aim *is* to draw selectively on the work of Foucault. Here, too, Weeks is concerned about the difficulty of incorporating Foucault into (conventional) historical research—for the reasons just outlined above. Nevertheless, Weeks's work is importantly shaped by four familiar Foucauldian themes. These are:

- the constructed aspects of discourse;
- the variability of the genealogies associated with such discourses;

- the emphasis on the details associated with the emergence of discourses; and
- the contestation associated with them.

In *Sex, Politics and Society* Weeks defends the view of the construction of homosexuality in the nineteenth century. The term 'homosexual' was first coined in 1869.[8] Of course, homosexual *behavior* existed before this date, but "The notion that 'a homosexual', whether male or female, could live a life fully organized around his sexual orientation is . . . of very recent origin."[9] It is in this sense that the claim is made that there were no homosexuals before the late nineteenth century.[10] Assuming an underlying model of variable psycho-sexual orientation, for Weeks the crucial *historical* question concerns:

> the emergence of the notion that homosexuality is a psychological or emotional condition peculiar to some people and not to others, and the social implications of this conceptualization.[11]

Of course homosexual behavior was severely regulated and punished prior to this period as well. But the transgressions which where prohibited and punished were not specifically a matter of homosexuality or even homosexual behavior. Sodomy, buggery and other "unnatural crimes" were, as it were, sex-, age- and even species-*neutral* offenses.[12] The emergence of specific prohibitions of homosexual acts did not occur until the later part of the nineteenth century—coincident with the emergence of the concept of the homosexual as a person, a concept that was shaped by the development of a medical notion of homosexuality as "characteristic of a particular type of person."[13]

Weeks embraces a Foucauldian view on the interplay of these various elements and the way in which they allowed the development of a "reverse discourse": "they created, in a variety of ways, self-concepts, meeting places, a language and style, and complex and varied modes of life."[14]

The problem here in thinking of this as a Foucauldian view, however, is the word "created"—for it is a causal word if ever there was one. The problem is not Weeks's. In fact, when he quotes Foucault, the relationship is one of what "made" what "possible":

> There is no question that the appearance in nineteenth century psychiatry, jurisprudence, and literature of a whole

> series of discourses . . . made possible the formation of a "reverse discourse" . . . .[15]

I will come back to the question of the causality implicit in the phrase "made possible" for Foucault himself later. For Weeks, at least, the history of the construction of homosexuality is shot through with causality, as is the rest of his history of the regulation of sex in Britain from the beginning of the nineteenth century onward. Forces are at work (that effect the family).[16] Family structures are created (by a variety of social practices).[17] Changing attitudes (toward birth control) result from other changing attitudes (toward children).[18] Working-class patterns of life are shaped (by values),[19] and conditions of emergence are proposed (for the regulation of sexual behavior).[20] That history is also one in which belief[21] and awareness[22] play a crucial role and, in the end, agency does as well.[23]

None of this should be surprising; it is just that none of it is regulation Foucauldian either. Causation, belief and agency play too central a role, and the genealogy of discourses about sex is too rooted in that role. That is to say, what Weeks does draw on Foucault for—the constructed aspects of discourse, the variability of the genealogies associated with such discourses, the emphasis on the details associated with the emergence of discourses, and the contestation associated with them—are all rendered in straightforward causal historical terms.

Weeks is not alone in this attempt to fit Foucault into causal history. The same move can be seen in some contemporary attempts to incorporate deconstruction into history as well. A good example is Mary Poovey's *Uneven Developments*.[24] Like Weeks, Poovey is interested in the history of the construction of sexuality in Victorian Britain and in the role of discourses in this construction. But, for her, the central organizing principle in this discourse is the *binary opposition* between the sexes. It is this binary opposition that Poovey argues "underwrote an entire system of institutional practices and conventions at mid-century, ranging from a sexual division of labor to a sexual division of economic and political rights."[25] So, for example, as competing interests fought for control in both obstetrics and nursing,[26] the binary discourse of sex served a central purpose[27] in the way in which issues were framed.

If power is one theme, individuals are another. For Poovey is also concerned with the way in which the development of discourses placed constraints on individual experience itself. This is no standard conception of binary opposition—at least in the sense of deconstruction. Here, there is no "reverse [of] binary oppositions and the notion of identity upon which it depends."[28] Nor is there any resulting substitution of "endless deferral or play. . . ."[29] Instead, Poovey's studies are historically rooted attempts to display the way in which the binary opposition of discourse about sex played a role in the organization of mid-Victorian life, and what is at stake in many of these studies are issues of power. Not power in the Foucauldian sense but in the old-fashioned conception of the "consolidation of bourgeois power."[30]

Like Weeks, Poovey is aware that she is a deviant when it comes to the use of post-structuralism in an historical context,[31] and it is the nature of this deviance that I now want to examine. For, if the use of post-structuralism in the *practice* of history is significantly different from the theory, then its import as an account of the construction of meaning in history may be different as well. But how deviant *is* the practice from the theory? The problem in answering this question is a function of how the theory itself is interpreted.

Prima facie, the problem Weeks and Poovey face is this: to shoehorn elements of the Foucauldian corpus into a standard conception of history is to ignore a central element of that corpus. And by parallel reasoning I think that it is self-evident that the same claim can be made with regard to other central elements of Foucault's thought with regard to language, knowledge and truth.

For the purposes of argument, I want to make a division of Foucault's theoretical claims into three distinct parts. This division can be thought of as purely arbitrary. It may turn out that we ought to be holists about Foucault and not make *any* such divisions of his work. Or it may turn out that I am not making a division in an enlightening way. But I want to sidestep these issues because I think they will turn out to be moot.

I also want to avoid the question of whether to treat Foucault's work as cumulative or not, by ignoring claims that Foucault's later work involved a repudiation of his earlier work. For my purposes it will be enough to maintain that there is continuity between *The Archaeology of Knowledge*[32] and *Discipline and*

Punish.[33] I focus on *The Archaeology of Knowledge* because it is the theoretical part of Foucault's work that has had by far the largest influence on historians. I will begin with truth, knowledge and language and only later circle back to matters of causation.

NINE FOUCAULDIAN THESES

Foucault's constructivism of discursive practices relies on two core commitments: *anti-essentialism* and *multiple determination*. For Foucault, anti-essentialism undermines the hope of settling the nature of things in an ahistorical way, and multiple determination underscores the prejudice that historical routes to the "same" phenomenon will not provide a source for unity either. To write a history of our *presentist* practices is on the one hand to immerse oneself in the minutiae of their *determinants*.[34] But Foucault's proposed method for establishing this is both *anti-causal* as well as *anti-phenomenological*.[35] Foucault's method also embraces three central philosophical theses about *truth, knowledge* and *language*. Speaking very much with the vulgar, and for now simply to list them: discursive practices are to be treated as non-referential; truth is to be treated as relativized to a practice; and knowledge is to be understood in non-Cartesian terms.

Of course, these are not all the Foucauldian theses one could list. But they are the ones that we need to attend to and I want to do so by dividing them into three different parts:

Division 1 anti-essentialism, multiple determination and presentism.
Division 2 the focus on etiology, acausality, and anti-phenomenology.
Division 3 the theories of truth, knowledge and language.

Division 1, taken by itself, is philosophically straightforward, even if it is contentious in other respects. I take those respects to be mainly a matter of history. Even if you are an anti-essentialist, to the extent that you want to avoid historicism, some generalization will be necessary, as we saw in the previous chapter. To write about homosexuality is already to write about a general category. That category may have no essential characteristics and its instances may have different sorts of determination. The challenge becomes, then, on what to base the notion of the category itself. One way to do this is to embrace presentism—to begin with our own cultural constructions.[36]

Now even if one accepts this program, and after all one need not treat it as an *exclusive* program,[37] major difficulties arise when you attempt to mesh it with the theses that make up Division 2. The theses of Division 2 wear their challenges for a "standard" approach (including that of Weeks) to history on their sleeves. First and foremost there is the question of how to reconcile the seeming contradiction between Foucault's interest in the pursuit of etiology and the claim of acausality. Of course, related to this is just how to make sense of the claim of multiple determination in Division 1 with the claim of acausality in Division 2. Finally, whether the account is to be causal or not, the third thesis in Division 2 seems to preclude building an account of origins that is tied to beliefs, let alone awareness and (depending on your theory of it) agency.

I will turn to the problems of interpretation associated with the theses of Division 2 later. Etiology and acausality will come to the fore later in this chapter and the problem of an alternative to the phenomenological is taken up in the next chapter. But before turning to these matters I want to concentrate on the theses of Division 3. For these theses not only challenge the very possibility of writing history as is "standardly" conceived, but a much broader swath of intellectual discourse as well.

We have already seen the way in which Foucault's theses about truth, language and knowledge are the theses at the core of the views of historians like Joan Scott. Whatever her practice, it is claims about language, truth and knowledge that prompt her to follow Foucault in proclaiming that "knowledge . . . [means] . . . the understanding produced by cultures and societies . . . ."[38] So that "it suggests that universal explanation is not, and never has been possible."[39] And that "knowledge is not absolute or true, but always relative."[40]

Thus, in this most conventional rendition, the theses of Division 3 form a tripartite attack on the idea of a correspondence account—be it of language, knowledge or truth, and there is much in Foucault's writing to give credence to this standard interpretation. Much of the *Archaeology of Knowledge* is structured around a defense of treating language in non-referential terms.[41] As such, the anti-realist implications for notions of both truth and knowledge follow straightforwardly, even though Foucault also provides independent reasons for holding to them as well.

At the most general programmatic level, an account of Foucault's views can be found in "The Art of Telling the Truth."[42] I quote at length because of the centrality of the position to the interpretations that follow:

> Kant seems to me to have founded the two great critical traditions between which modern philosophy is divided. Let us say that in his great critical work Kant laid the foundations for the tradition of philosophy that poses the question of the conditions in which true knowledge is possible and, on that basis, it may be said that a whole stretch of modern philosophy from the nineteenth century has been presented, developed as the analytics of truth.
>
> But there is also in modern and contemporary philosophy another type of question, another kind of critical interrogation: it is the one we see emerging precisely in the question of the *Aufklärung* or in the text of the Revolution. That other critical tradition poses the question: What is our present? What is the present field of possible experiences? This is not an analytics of truth; it will concern what might be called an ontology of the present, an ontology of ourselves, and it seems to me that the philosophical choice confronting us today is this: one may opt for a critical philosophy that will present itself as an analytical philosophy of truth in general, or one may opt for a critical thought that will take the form of an ontology of ourselves, an ontology of the present; it is this form of philosophy that, from Hegel, through Nietzsche and Max Weber, to the Frankfurt School, has founded a form of reflection in which I have tried to work.[43]

This important statement marries Foucault's presentism to his views about the role of philosophy. But, if the above can be read as supportive of the kinds of views championed by Scott, in fact it contains a crucial ambiguity that allows *two* competing interpretations of how the analytic tradition is to be viewed. On one interpretation, call it the weak one, the choice Foucault speaks of is a *real* choice. On the other, call it the strong interpretation, there is only one alternative. What is at stake is whether or not the notion of an "analytic philosophy of truth in general" is a program we *could* pursue but choose not to (the weak interpretation) or whether in fact this choice is a chimera (the strong interpretation).

It will turn out that there is a parallel to this distinction that can be found at a more restricted level as well—namely, how we should understand the *scope* of Foucault's epistemological project. If *The Order of Things* is intended as a history of the human sciences, *The Archaeology of Knowledge* is more ambiguous in its ambition. Should we understand "knowledge" here to be *all* knowledge (the strong interpretation) or knowledge as restricted to the human sciences (the weak interpretation)?

I call these contrasts (at the restricted level) "strong" versus "weak" not just because of the issue of their scope but also because of their philosophical implications. If you take the strong view, the status of your own knowledge claims, truth claims and discourse exposes itself to the danger of being undermined by the act of self-reflexive application.[44] Thus, writing in her introduction to *Uneven Developments,* Mary Poovey issues the following caveat about her work:

> To argue that knowledge is socially constructed, as I do in this book, is necessarily to admit that one's own interpretations are part of larger social constructs and, as a necessary corollary, that they are ideological . . . to adopt the position that I have adopted is to renounce even the pretense of objectivity.[45]

Well, maybe. But before embracing this view and its philosophical consequences, we should ask whether it should be viewed plausibly as part and parcel of the Foucauldian corpus. Yet, whatever the answer to that question turns out to be, in the end we will have to examine the view on its own merits, irrespective of whatever the most plausible Foucauldian canon on the matter turns out to be. (I should note that by the "most plausible Foucauldian canon," I do not simply mean plausible by Foucault's lights. The issue is not one of authorial intent as it refers to the scope of the argument. By that standard there will be little debate about the strong versus the weak interpretation—as Bruno Latour has emphasized to me,[46] Foucault never took himself to be writing about the natural sciences and hence about all of knowledge. He, and nearly all of his French contemporaries, took the debate to be about the human sciences. But I take the "plausible canon" to be more than just a matter of Foucault's own actual intended scope for his argument. Especially in the Anglo–American context, the desire to see Foucault's work in the broadest possible context has inspired the quite natural interpretation that natural science, and with it all of knowledge, ought not to be offered some special protective status. As such, in what follows, when I speak of "the intended scope of Foucault's argument" I will use the phrase in the sense of what may be inferred from Foucault's work itself, unconstrained by any independent notion of authorial intent. In doing so, I will also make another assumption; namely, that in matters of interpretation, irrespective of authorial intent, consistency is a virtue.)

INTERPRETATIONS

A vast and varied literature exists covering all aspects of Foucault's work. Still, if one approaches that literature with the narrow philosophical focus that interests me here, it is possible to classify it into a number of standard "moves." On the "strong interpretation," what is at issue is the status of Foucault's own statements *about* these matters, given that (on the strong interpretation) these statements fall within the domain of these same matters themselves.

A number of commentators, including Merquior[47] and Putnam,[48] have developed the argument that Foucault's work should be accorded the strong interpretation and that, hence, it is self-refuting.[49]

One response to this line of reasoning might be to argue as follows: a plausible account of Foucault's intended scope for his argument ought to be guided by an interest in not undermining the argument itself. Yet the strong argument as is does undermine the argument. So it can't be the correct interpretation.[50] However, there is another argument for the strong interpretation that is consistent with refusing to accept the condition that a plausible account of Foucault's intended scope for his argument ought to be guided by an interest in not undermining the argument itself. That alternative is to argue that the structure of Foucault's argument is not one that forces him to be concerned with its vulnerability to being undermined on its own terms. A second "family" of commentators has developed this line of reasoning by stressing the elements of Nietzscheanism to be found in Foucault's philosophy.

It is certainly indisputable that Foucault thinks of himself in this tradition explicitly in "Nietzsche, Genealogy, History"[51] and implicitly in *The Archaeology of Knowledge*. Furthermore, Foucault's identification with this tradition influences the conception of history that he takes himself to be criticizing as the "pursuit of origin." Pursuit of origin here is not causal; rather "it is an attempt to capture the exact essence of things . . . this search assumes the existence of immobile forms that precede the world of accident and succession."[52]

Having said this, what is at issue is just how much of a Nietzschean is Foucault? Is he the kind of Nietzschean about whom Richard Rorty asserts: "The Nietzschean wants to *abandon* the striving for objectivity and the intuition that Truth is One, not to redescribe it or to ground it."[53] Rorty is certainly not alone in thinking that he is. But then how should we interpret Foucault's methodological writing, especially in *The Archaeology of Knowledge*? Both Alan Megill[54] and Larry Shiner[55] counsel against taking it seriously *qua* a method. Instead, they argue that it is a deliberate parody of method that serves a rhetorical purpose of criticizing it.

The virtue of this line of interpretation is that it does achieve the goal of avoiding the undermining effect of the strong interpretation that occurs when it is taken on face value. But there is a price to be paid for this move as well. To be plausible, any interpretation of Foucault's methodological writing has to be rendered consistent with Foucault's substantive (non-methodological) writing. And, by viewing the methodological writing as a parody, it is not easy to see just how this consistency is to be achieved, given the evident seriousness with which Foucault took his substantive writing. That is not to say that seriousness and parody cannot be rendered compatible. But it is hard to read the descriptions of torture in *Discipline and Punish* and not take them at face value.[56] If we do that, then how are the substantive and the methodological to be reconciled? *Prima facie*, the two cannot simply be allowed to coexist—one taken as parody, and the other at face value. For, if the methodological is to be treated as a parody of method, it is a parody that is meant to carry a message. The message is meant to undermine the epistemic status of historical claims among others. But then how do we prevent this message from also undermining the epistemic status of Foucault's own substantive work?

The way I have just put this dilemma assumes that seriousness is only achieved by way of an epistemic route that is itself vulnerable to Foucault's methodological critique. It also assumes that the only choice is between parody on the one hand and seriousness so understood. Of course, it ain't necessarily so. But it *is* so on the interpretation that we have been considering. A counter-interpretation can push from a number of different directions. One is against the view that Foucault's arguments are parodic. Another is against the view that the substantive work relies on epistemic assumptions that are vulnerable to Foucault's own critique.

Still, these alternatives notwithstanding, there is a more general lesson to be learned from the discussion of the view of Foucault as anti-methodologist. It is that any successful interpretation of Foucault's views must not only deal with the question of method but also of its relationship to Foucault's substantive work. As such we can certainly conclude at least this: the view that holds that Foucault is offering an anti-method does so by complicating the interpretation of the substantive work, even if it solves the problem of the undermining force of self-reference.

But I think that there is a more straightforward reason to reject the view that Foucault is offering an anti-method as well: if Foucault takes torture too seriously in *Discipline and Punish* to be seen as a parodist in his work on methodology, he takes truth and reference too seriously in *The Archaeology of Knowledge* and the "Discourse on Language" to be understood merely as offering an anti-method—at least an anti-method *simpliciter*. I will have more to say about the substance of these views in a moment, but for now, it is important to realize just how carefully Foucault circumscribes his project with a view to leaving a conception of truth and reference undisturbed within other domains of application. Thus, in *The Archaeology*, sentences *do* have referential capacity:

> A sentence cannot be non-significant; it refers to something by virtue of the fact that it is a statement.[57]

And that reference is not subject to endless play:

> what might be defined as the *correlate* of the statement is a

> group of domains in which such objects may appear and to which such relations may be assigned . . . .[58]

even if their reference is holistically determined.[59] As with reference, there is a place for truth as well:

> It is always possible one could speak the truth in a void . . . .[60]

Further:

> the division between true and false is neither arbitrary, nor modifiable, nor institutional, nor violent.[61]

The expression of sentiments like these lend credence to the view that Foucault's project is much less global in ambition than the interpretations considered so far take for granted. Indeed, in places, Foucault is explicit about the limits of his project and how it can *coexist* with others: "the analysis of statements does not claim to be a total, exhaustive description of 'language' (*langage*), or of 'what was said'."[62]

Yet how can this be, on what I have termed the "strong interpretation"? For on that view everything *does* fall within the purview of the analysis. Then so much the worse for the strong interpretation—for, after all, it is the source of the problems of self-reference. But can the weak interpretation that avoids these problems be sustained?

The weak interpretation is that Foucault's project is limited to domains that do not include his own discourse, so that the problems of self-reference are avoided. The challenge here is that, on the one hand, a "cut" has to be defended between Foucault's subject and his own discourse. Doing that pulls in the direction of interpreting the subject narrowly, thus maximizing the chances of leaving his discourse outside of the subject. But if the subject is drawn *too* narrowly, the power of the interpretation will be undermined by its limited range of application. Furthermore, aside from the problem of power, perhaps there is *no* cut that can be drawn, however narrowly, that will allow the distinction bruited above to be drawn—that is to say, maybe Foucault's discourse is implicated in all domains.

Certainly Gary Gutting thinks that there is such a cut to be made. His *Michel Foucault's Archaeology of Scientific Reason* is an important and articulate defense of the weak interpretation.[63]

But I think that Gutting's line of defense fails. For Gutting, the strategy is to draw a line around the "human sciences" as Foucault's arena of interest. Gutting wants to defend the distinction between Foucault's views about truth and objectivity in these disciplines and those that do not involve human beings:

> There is no suggestion that he [Foucault] thinks his archaeological method could be applied to sciences like physics or chemistry to show that their claims to truth and objectivity are questionable.[64]

Let us suppose we accept this for the purposes of argument, and along with it "the local or regional nature of his analyses."[65] Still, how does this avoid Foucault's own discourse from falling within the domain he discusses? Foucault does speak of mathematics as if it had a special (and protected) epistemological status[66] and one might try to extend this to logic, and perhaps thereby to a notion of justification. Yet, at best, what we end up with is a concept of philosophy that is enough for the Vienna Circle but not for Foucault's broad intellectual brush.

Gutting conceives of the problem as one of separating different subject matters: the hard sciences versus the rest. But to protect Foucault's own theoretical discourse, more terrain needs to be treated as beyond its reach. Yet, given the content of that discourse, it is not clear that such a strategy will work.

I think that there is a different way to see what is going on in Foucault that does allow us to draw the line between his subject matter, while protecting his own discourse. It is not so much a matter of the subject matter as a matter of the *approach* to the subject.

TRUTH AND IN THE TRUTH

Though *The Archaeology of Knowledge* may end up *inadvertently* constituting a global argument, in fact, it is not so as a matter of design. For Foucault is explicit about the way in which the normal referential uses of language need to be set aside in order to allow a focus on its discursive features. It is not that the referential features of language do not, or cannot, obtain. It is just that their presence obscures other features of language:

> the analysis of statements does not claim to be a total, exhaustive description of "language" (*langage*), or of "what was said."[67]

Indeed:

> the "signifying" structure of language (*langage*) always refers back to something else; objects are designated by it; meaning is intended by it .... Language always seems to be inhabited by the other, the elsewhere, the distant.... But if one wishes to describe the enunciative level, one must consider that existence itself; question language, not in the direction to which it refers, but in the dimension that gives it [sic]; ignore its power to designate, to name, to show, to reveal, to be the place of meaning or truth ....[68]

But if the referential function can be temporarily suspended for the purposes of inquiry, is the same true of truth? Perhaps we can ignore language as the place of truth, but what of truth itself? Here, too, Foucault's agenda is much less global than many have thought. For there is in Foucault a notion of truth that he wants to preserve as stable and relatively straightforward whatever else is going on. What "The Discourse on Language" makes clear is the contrast between the *will to truth* and being "in the true" (*dans le vrai*) by means of showing how both are different from the truth *simpliciter*.

The first contrast is a straightforward one that poses little philosophical challenge. Our will to truth can be understood as our epistemic interests—"our will to knowledge."[69] That involves a historical construction through which it is constituted as a separate interest from other interests,[70] and yet the constructed nature of these interests notwithstanding, they leave truth *simpliciter* unaffected: "the division between true and false is neither arbitrary, nor modifiable, nor institutional, nor violent."[71] However, things get much more complicated when it comes to the contrast between truth *simpliciter* and being "in the true."

To be "in the true" is to be within the discursive practice of a discipline which is constitutive of that discipline itself:

> a proposition must fulfill some onerous and complex conditions before it can be admitted within a discipline; before it can be pronounced true or false it must be, as Monsieur Canguilhem might say, "within the true."[72]

It is tempting to read this in Kuhnian terms. After all, Foucault does think of the "well-defined process associated with disciplines as a sine qua non for a notion of error to make any sense at all,"[73] and he also thinks that to be recognized as true requires conformity to the conceptual apparatus of a disciplinary context. Still, having said all of this, Foucault also believes in a notion of truth that is context- (discipline-) independent. True, to understand disciplines, you have to study what constitutes being "in the true," but for Foucault there is also the truth *simpliciter*, and, unlike Kuhn, the truth *simpliciter* can obtain in a context-independent way: "It is always possible one could speak the truth in a void . . . ,"[74] even if, at the same time, the concepts one uses don't so exist, since, "without belonging to any discipline, a proposition is obliged to utilize conceptual instruments and techniques of a well-defined type . . . ."[75]

This last claim is an important caveat. Concepts are not given or under the control of individuals. Here, Foucault is a globalist, embracing the strong interpretation with respect to concepts, and it is natural to argue that truth itself is a concept and so that it, too, falls under these global claims. "Truth," then, is a constructed notion with a particular history, and so on. But in fact this is not the way Foucault argues. In fact he wants to leave truth *simpliciter* out of the picture and follow his interests in the process by which truth is "pronounced" as such within particular discursive practices. It is not truth, but what counts as truth, that matters to him, and what counts as viewed within particular disciplinary traditions. But is Foucault entitled to such a distinction in the first place? Can there be a notion of truth independent of the means by which we declare propositions as true? Let us be realists about truth for a moment and allow that there is a fact of the matter about whether a proposition is true or not. Let us also be pragmatists about the notion of truth as well, and allow what is true to be a function of a communal agreement. Now let these two conceptions of truth coexist. This level of tolerance is alien to most realists and pragmatists, for they would only allow for one conception of truth. But Foucault's pragmatism here does not aspire to a full-blown account of truth; rather, it restricts itself to the disciplinary process by which truth is "pronounced." Still, one might object as follows: why should the realist concept of truth be exempted from the globalism that attaches to all other concepts and with it the strong interpretation? Here is what strikes me as a plausible gloss on how Foucault might be saved from this objection. To "speak the truth" does not require us to appeal to the concept of truth itself. If we embrace eliminativism about truth, we speak the truth in asserting that "snow is white" simply if snow is white.

If there is a notion of truth to be had independent of what counts as truth, does this help sidestep the problem of the self-refuting aspect of Foucault's argument? At a first glance the answer would seem to be "no," for the following reason: all writing constitutes part of a discursive practice and is hence subject to the rules of that practice, and so on. And so, too, with Foucault's writing. How can Foucault's historical and philosophical writings avoid considerations of truth as a matter of those disciplines and fall into the domain of truth *simpliciter*? The answer is that they cannot. But that does not mean that they cannot do both. That is to say, there is nothing in Foucault's argument to prevent any of our discourse from being both true and "in the true" with respect to one or more discursive practices at the same time. And that gives us and Foucault a leg on which to stand.[76]

Like Gutting, then, I think that the weak interpretation is the correct one. Foucault's scope is not global. But I think Gutting cuts the world up in the wrong way. It is not a matter of disciplines, but a matter of the approach to those disciplines that is at stake. That is to say, of any discipline we can ask questions about its truth. But on the weak interpretation those questions will not infect the status of our own discourse *qua* its truth *simpliciter*.[77]

FOUCAULT AND HISTORIANS

If the weak interpretation is best suited as an interpretation of Foucault on internal grounds, how does this interpretation accord with how historians have made use of Foucault? Here, as before, we need to distinguish between theory and practice. For, as we saw earlier, the claims of historians writing theory are different from historical practice. We now have a taxonomy on which to try to hang this difference. It is this: while the claims of theory are those of the strong interpretation, historians' practice is much more in the spirit of the weak interpretation.

I take it that this claim about the strong interpretation is not very interesting. After all, the historians discussed earlier who have written about Foucault have done so in the argot of a received post-structuralist tradition that takes the strong interpretation for granted. It is the claim about the weak interpretation that needs defense.

The difficulty with the line of reasoning that I am pursuing here is that what I have termed "historians' practice" involves much more than what I have just attributed to the weak interpretation: namely, the possibility of maintaining working use of both reference and truth *simpliciter*. That much *is* certainly in play in the practice of historians. But we saw more as well. Recall Weeks. I argued earlier that causation, belief and agency played too central a role in his account of the construction of homosexuality to count as standardly Foucauldian, and, as a result, I suggested that Weeks was forced to fit Foucault into a standard conception of history. Does embracing the weak interpretation change any of this? As things stand, it cannot. For the weak interpretation has had nothing to say about causation, belief or agency, and *prima facie* extending it to say something about these subjects will be a hard row to hoe, given Foucault's explicitly non-"standard" views on these matters. Yet Foucault's anti-causalism has to be carefully tempered by the recognition that, in non-technical uses, "causality" has two meanings that are commonly conflated. One is a matter of determinism, in the broad sense that we associate with the claim that there are laws of history. The other is a matter of whether particulars are amenable to an analysis in terms of their etiology. Rejection of the first of these two senses of causality is a central tenet that drives much of the Foucauldian program: "we must rid ourselves of a whole mass of notions, each of which, in its own way, diversifies the theme of continuity."[78] And that includes "the notion of influence . . . which refers to an apparently causal process . . . of resemblance and repetition . . . ."[79]

But Foucault takes a much more benign attitude toward the second notion. Causation, in this sense, like reference, is something that obtains. It is just that Foucault's interests pull him in a different direction and that direction is one that he thinks is ultimately more fundamental and prior to causal analysis. Let me quote at length from a crucial statement of the way in which Foucault sees this contrast:

A causal analysis . . . would try to discover to what extent

> political changes, or economic progress, could determine the consciousness of scientists. . . .
>
> Archaeology situates its analysis at another level: the phenomena of expression, reflexions, and symbolization are for it merely the effects of an overall reading. . . . The field of relations that characterizes a discursive formation is the locus in which symbolizations and effects may be perceived, situated, and determined. If archaeology brings . . . discourse closer to a number of practices, it is in order to discover far less "immediate" relations than expression, but far more direct relations than those of causality communicated through the consciousness of speaking subjects.[80]

Here, belief and its causal determinants are treated as legitimate objects of historical interest—it is just that they are less interesting than the processes revealed by an archaeological approach.

So here we have an allowable sense of causation, and, in fact, it is the relevant sense of causation when one turns to look at the practice of historians. Weeks is surely typical in having no interest in a notion of historical law but in the causal determinants of particular events and phenomena. The only problem is that the phenomenon in which he is interested is the emergence of discourses about sexuality and their causal determinants. But this looks like much more than is allowable *even* on the weak Foucauldian interpretation. For, on the version of the weak interpretation that I have been defending, causation can be thought of like the referential features of language. As the above quotation indicates, it is not ruled out as impossible. But it is nonetheless something to be set aside as we examine discursive practices.[81] Yet what this rules out is pursuing causal questions about discursive practices—including those about sexuality.

FOUCAULDIAN THEORY AND PRACTICE

The opposition in which Foucault places the archaeological method and causal considerations is, of course, far reaching. It is a central thematic element of *The Archaeology*:

> Discourse . . . is not an ideal, timeless form that also possesses a history; the problem is not therefore to ask oneself how and

> why it was able to emerge and become embodied; it is, from beginning to end, historical—a fragment of history, a unity and discontinuity in history itself, posing the problem of its own limits, its divisions, its transformations, the specific modes of its temporality rather than its sudden irruption in the midst of the complications of time.[82]

Yet, what is wrong with asking oneself how and why discourses were able to emerge and become embodied *at these various points in time*? Of course the problem here is that Foucault's argument is directed against a notion of discursive formations as historically stable formations. Yet, when causation (or causality) is understood in the sense of the determinants of events and processes rather than a matter of historical laws, this opposition is hard to sustain—even for Foucault—in historical practice.

The argument here runs parallel to a consideration raised earlier against the interpretation of Foucault as a parodist: namely, Foucault takes torture too seriously in *Discipline and Punish*. In fact, the changing "discourses" of punishment that are the central focus of *Discipline and Punish* are embedded in a causal nexus of both etiologies and effects. Thus "The art of punishing . . . must rest on a whole technology of representation"[83] to effect a ritual recoding of crime *in the society*. So, for example:

> How can one silence the dubious glory of the criminal? This was a matter of grave concern to the lawmakers of the eighteenth century. . . . If the recoding of punishment is well done, if the ceremony of mourning takes place as it should, the crime can no longer appear as anything but a misfortune and the criminal as an enemy who must be re-educated into social life. . . . Discourse will become the vehicle of the law: the constant principal of universal recoding.[84]

Here at least there is no tension between historically specific discourse and historically specific causation.

Can we generalize this as a part of what I have termed the "weak interpretation"? I want to remain agnostic about whether or not it is plausible to say that this expanded version of the weak interpretation is an *echt* Foucauldian position. Instead, I prefer to leave it as the position that one ought to hold—for there is no incompatibility between causation (in the sense we have been using it) and discursive practices that justifies ruling it out. Let us call this interpretation prescriptive as well as weak. Prescriptive or not, I think that it is the position that is most congenial with historical practice that operates with a Foucauldian influence and that includes Foucault himself. Allow also for the purposes of argument that considerations of causation do not, in themselves, support an anti-realist position.[85] Then there is a philosophical conclusion that follows from the fact that there is no incompatibility between the study of causation and the study of discursive practices: Foucauldian-inspired history does not in itself require an anti-realist view of history.

Paul Veyne, probably the most insightful commentator on Foucault, has written that Foucault's central and most original thesis was that:

> *ce qui est fait,* l'object, s'explique par ce qu'a été le *faire* à chaque moment de l'histoire; c'est à tort que nous nous imaginons que le *faire,* la pratique, s'explique à partir de ce qui est fait.[86]

I think Veyne is correct in his assessment. Foucault's great contribution was his insistent reminder against our temptation to reify our practice and see categories as though they were given in nature. But I have tried to show that this Kantian *aperçu* is compatible with the notion of tracing the causal import of our practice. To speak of the categories of experience as constructed is not to say that we cannot ask questions about the circumstances of that construction. Nor that we cannot answer these questions in terms about which there is a fact of the matter.

THE DERRIDEAN EXTENSION

Over the last two chapters I have been arguing that, while you can pull what is an anti-realist view of history out of Foucault's theoretical writings, the position ends up as hard to sustain. The practice of historians, Foucault's own historical practice and, in the end, a plausible reading of his theory all militate against the view.

At some points in this argument I have freely helped myself to the work of historians influenced by Derrida as well as by Foucault. But I now want to try systematically to canvass the prospects for extending the first of these three considerations (the practice of historians) to cover work inspired by Derrida.

I have been arguing that historians' assimilation of Foucault's work is in fact consistent with what I consider to be the most plausible reading of his theory as well as his own practice. But when it comes to the influence of Derrida things get more complicated. There is only a partial parallel to the line of reasoning that I have been using about the influence of the work of Foucault.[87]

The general thesis that I want to illustrate is how selective historians have been in the part of the Derridean corpus that they have assimilated and how that part will not support an anti-realist view of history any more than the use of the work of Foucault. What drives this selectivity is that Derridean views *taken as a whole* are incompatible with the programs of such historians. But here, as in the case of Foucault, I have in mind the *practice* of historians not their theoretical conceptions of what it is they take themselves to be doing.

Unlike the "strong interpretation" of Foucault that I disfavored over the "weak interpretation" as the correct interpretation of Foucault, in the case of Derrida I think that the opposite is true. In Derrida's own work there is no avoiding what I will here, too, call the "strong interpretation."

Jonathan Culler would not agree. In *On Deconstruction*[88] he reminds us of an important passage from Derrida's essay "Structure, Sign and Play in the Discourse of the Human Sciences":

> There are . . . two interpretations of interpretation, of structure, of sign, of play. The one seeks to decipher, dreams of deciphering a truth or an origin which escapes play and the order of the sign, and which lives the necessity of interpretation as an exile. The other, which is no longer turned toward the origin, affirms play and tries to pass beyond man . . . [who] has dreamed of full presence, the reassuring foundation, the origin and the end of play . . . .
>
> For my part, although these two interpretations must acknowledge and accentuate their difference and define their irreducibility, I do not believe that today there is any

> question of *choosing*—in the first place because here we are in a region (let us say, provisionally, a region of historicity) where the category of choice seems particularly trivial; and in the second, because we must first try to conceive of the common ground, and the *différance* of this irreducible difference.[89]

For Culler this passage underscores Derrida's commitment to a view of interpretation that is something more than just the second interpretation alone—that is to the celebration of play:

> Derrida has often been read as urging us to choose the second interpretation of interpretation, to affirm the play of meaning; but as he notes here, one cannot simply or effectively choose to make meaning either the original meaning of an author or the creative experience of the reader.[90]

Instead we are "stuck" with a "combination of context-bound meaning and boundless context . . . ."[91] As Culler and others have pointed out, it is certainly true that Anglo–American philosophers have been exclusively interested (and worried) about the implications of the second interpretation.[92] For it is the second interpretation that gives the Derridean approach its metaphysical force—it is the notions of play, bricolage, and différance, and the critique of presence[93] that all undermine any hope of fixing meaning by way of speaker or authorial intentions. Yet, as John Ellis argues in *Against Deconstruction*, Culler's gloss on Derrida does nothing to change this picture. For combining the two interpretations simply infects what is at least the *prima facie* determinacy of the first interpretation with the indeterminacy of the second.[94] Hence, by the unavoidability of what I have termed the "strong interpretation," I simply mean that there is no plausible interpretation available of the Derridean version of the argument that avoids the metaphysical embrace that the deferral of reference entails.[95]

Gyan Prakash is surely correct when he chastises his critics for claiming that "Derrida's critique of foundational thought 'can do little more than reveal, over and over again, the subjective and arbitrary nature of our categories and the uncertainty of knowledge derived from them'. "[96] For Prakash:

> Deconstruction is emphatically not about showing the arbitrariness of our categories; rather its purpose has been to

> show that structures of signification effect their closures through a strategy of opposition and hierarchization that edit, suppress, marginalize everything that upsets founding values. And yet the very staging of this strategy reveals what is repressed.[97]

But if deconstruction is emphatically not about showing the arbitrariness of our categories, is it simply all that follows in the above quotation? I say "simply" in the sense that what follows in the above quotation holds no philosophical mystery. It champions a set of analytic techniques that allows voice to be given to that which is suppressed. If I am right that the strong interpretation of deconstruction is unavoidable, then Prakash's appeal to analytic technique cannot count as *echt* deconstruction. Derrida's two interpretations of interpretation cannot be so easily disentangled. But it seems to me that Prakash's characterization is precisely correct when it comes to the practice of historians who make use of Derridean notions. I want to illustrate this point by studying the work of the so-called "Subaltern Studies" Group, whose research concentrates on colonial India.[98] Whether we call it *echt* deconstruction or not, I want to argue that, given their interests, the Subaltern Studies Group's work cannot be viewed as supporting the anti-realism that flows from the strong interpretation. I want to contrast this to the theoretical claims made *about* the Subaltern Studies Group by Gayatri Spivak.

What is at issue in the meta-discussions of Subaltern Studies and other historical research programs that make use of deconstructive techniques is the issue of grounding. Nowhere is this more contentious than in the issue of peasant consciousness. As Gayatri Spivak puts it:

> To investigate, discover, and establish a subaltern or peasant consciousness seems at first to be a positivistic project—a project which assumes that, if properly prosecuted, it will lead to firm ground, to some *thing* that can be disclosed.[99]

The emphasis should be on "seems" here, because, of course, Spivak herself thinks that we should not be deceived by first appearances. Instead:

> the subaltern's view, will, presence, can be no more than a theoretical fiction to entitle the project of reading. . . . I

could put it thus: "Thought [here, the thought of subaltern consciousness] is here for me a perfectly neutral name, the blank part of the text, the necessarily indeterminate index of a future epoch of difference."[100]

There certainly is no avoiding the notion of peasant consciousness as a central theme of Subaltern Studies. The declared program of the Subaltern Studies Group[101] is to engage in historical research on the role of the people (the subalterns) in the development of Indian nationalism, in contrast to the roles played by both the indigenous elites as well as the colonizing power.[102] Writing such a history depends on documentation and archival material, nearly all of which was the product of the colonial infrastructure. Providing a reading of these materials with the role of the people in mind *is* extraordinarily well suited for the application of deconstructive techniques. It is just the kind of material which must be read for its silences, for what it tries to "edit, suppress and marginalize,"[103] and in all of this it is the rebel voice that is the object of the search.[104] Thus, Sahid Amin writes of:

> the displacement of the rebel in the judicial discourse . . . silences of the peasant accused are superimposed upon and given meaning by the instrumental "speech" of the rebel turned Queen's Evidence . . . .[105]

The judicial is only one "site" in which this process takes place, of course. For the historian, both the contents of archives and historiography itself need similarly special "readings" in attempts to constitute a rebel history.[106]

But if deconstructive techniques can be successfully used in the construction of a rebel voice from the "official" sources, we have to ask what it is that *constitutes* a voice. I think the answer that comes from a reading of the Subaltern Studies Group goes to the heart of the issue of groundedness. The rebel voice in these studies is not an abstract entity but an historically situated (often religious)[107] reality that was a causal agent in the (failed) rebellions against the Raj, and a part of providing an account of that agency means taking account of rebel consciousness.[108] That is to say, providing a "reading" of the archival text to give voice to the subaltern requires a hermeneutic turn that is attentive to rebel attitudes,[109] perceptions,[110] conceptualizations,[111] consciousness,[112] hopes,[113] intentions,[114] experience,[115] expectations,[116] perspectives,[117] motives,[118] world-views,[119] and beliefs.[120] The view is best summed up by Guha: "Insurgency, in other words, was a motivated and conscious undertaking on the part of the rural masses."[121]

Spivak herself is anxious to contrast a notion of consciousness that is historically specific with a "general" notion of consciousness.[122] This is an important distinction, for, to paraphrase Spivak, the historical specificity of consciousness contradicts the notion of general consciousness as indivisible.[123] But what about the role of historically specific consciousness? Granted, it can't be "*the* ground that makes all disclosures possible."[124] But that is not to say it cannot be a consideration, at least in this sense: it provides a favored "reading" over other alternatives. To put the matter in the argot of realism: there is a fact of the matter when it comes to historically specific consciousness, given its centrality to insurgency.

Now if this view of consciousness is right, the problem of compatibility with *echt* Derrideanism follows directly. For, as we have already seen, speaking of historically specific consciousness, Spivak thinks that "the subaltern's view, will, presence, can be no more than a theoretical fiction to entitle the project of reading."[125]

Not surprisingly then, for Spivak, there is an interest in rendering the historiographic construct of "subaltern consciousness" as an artifact of something else. She offers two hypotheses. One is that subaltern consciousness involves the substitution of "an effect for a cause":

> that which seems to operate as a subject may be part of an immense discontinuous network ("text" in the general sense) of strands . . . determined by heterogeneous determinations which are themselves dependent upon a myriad of circumstances, [and] produce the effect of an operating subject.[126]

The other is that, in the Subaltern Studies Group, subaltern consciousness is a case of "a *strategic* use of positivist essentialism in a scrupulously visible political interest."[127]

But notice that, if either of these views is correct, we lose the ability to provide an account of insurgency as "a motivated and conscious undertaking on the part of the rural masses."[128] For, once we bracket our realist assumptions about consciousness, then our ability to appeal to it as an historically specific lever of action is lost as well, and with that we lose most of its explanatory power. Ironically, there is a poignant—if inadvertent—illustration of just this point to be had in Spivak's own work:

> A young woman of sixteen or seventeen, Bhuvaneswari Bhaduri, hanged herself in her father's modest apartment in North Calcutta in 1926. The suicide was a puzzle since, as Bhuvaneswari was menstruating at the time, it was clearly not a case of illicit pregnancy. Nearly a decade later, it was discovered that she was a member of one of the many groups involved in the armed struggle for Indian independence. She had finally been entrusted with a political assassination. Unable to confront the task and yet aware of the political need for trust, she killed herself.
>
> Bhuvaneswari had known that her death would be diagnosed as the outcome of illegitimate passion. She had therefore waited for the onset of menstruation. While waiting, Bhuvaneswari . . . perhaps rewrote the social text of *sati*-suicide in an interventionist way. . . . She generalized the sanctioned motive for female suicide by taking immense trouble to displace (not merely deny), in the physiological inscription of her body, its imprisonment with legitimate passion by a single male. . . . The displacing gesture—waiting for menstruation—is at first a reversal of the interdict against a menstruating widow's right to immolate herself.[129]

5

EXTENDING THE ARGUMENT

In the last three chapters I have been examining philosophies of history against the backdrop of the practice of history. The argument strategy I have relied on is a demonstration of how the commitments of practice serve to undermine the commitments of theory by presuming a realist style of reasoning, given that our historical interests include the experiential and the causal.

The experiential and the causal have operated as twin, independent supports for the argument. In Chapter One I argued that, even if you discount experience in favor of the social construction of the categories of experience, "construction" draws on a causal metaphor—one to which I have suggested we bring a realist style of reasoning. And in the subsequent chapters we have repeatedly seen the way in which taking seriously the notion of the "horizon of possibility" of historical actors and detailing the contestation for control of culture call forth this style of reasoning. Historical understanding of the point of view of German Jews or Indian peasants drew on the first pillar. Historical understanding of the discourse of the family and the institution of punishment drew on the second. There is no profound step of philosophical reasoning from these illustrations of experience and culture to realism—instead, my claim is that our process of understanding appeals to assumptions we make about our own world. I am going to come back to consider some objections to these claims in the concluding chapter of this book, but in the present chapter I want to try to extend the argument in one particular direction that has been implicit in some of the preceding illustrations. I think that it is hard to make sense of the notion of experience without bringing individuals into the picture.[1] Be that as it may, how far can we go without individuals? In particular, if our interest is in the notion of the *interpreted* world, can we make sense of the social construction of *such* worlds as *prior* to experience?

To speak of eschewing experience,[2] while speaking of the construction of culture out of which the categories of experience are drawn, presumes that it makes sense to speak of the interpreted world prior to experience. It assumes that we can make sense of a notion of meaning that is, as it were, not in the head, and it assumes that such a notion can be implicated in causal claims to which we bring a realist style of reasoning.

What does it mean to speak of meaning as not in the head, and can such a notion be implicated in causal analysis? We can speak of Weberian representative individuals or ideal types of individuals,[3] and these remain true to the spirit of the analysis we have been pursuing. Yet to speak of cultural analysis ought not to require that we stay so close to the ground. But in what sense can we speak of meaning and interpretation once our feet are off the ground? Michel Foucault's way out of this dilemma was to ground such questions on practice.[4] Yet "practice" only starts us along the way. To speak of "practice" as a starting point instead of "experience" is all well and good. The problem is that the notion of what things mean makes sense when it is grounded in experience. But what does it mean when we leave the comfort of the phenomenological? This problem is in part epistemological, but as we shall see it is predominantly ontological in nature.

Habermas asserts that:

> The empirical context of actions regulated by social norms transcends the manifest meaning of intentions and calls for an objective frame of reference in which the latent meaning . . . can be grasped . . . .[5]

But just how are we to understand the conjunction between the objective and the latent?

Of course it is Clifford Geertz who is rightly taken as having "trumpeted . . . the study of culture as embodied in symbols," thus removing "the problem of getting inside people's heads . . . ."[6] His influence on writing cultural and social history over the last twenty years has been enormous. I think there are many versions of what has come to be thought of as a "Geertzian" approach. At least one of these is strongly anti-realist. In this chapter I want to examine this strain of anti-realism. My aim is to disambiguate it from other Geertzian interpretations and, in the end, use it as a foil to provide an outline of an account that "fits" what we want—meaning as not in the head but causally implicated. But, in the end, I will argue that such an account cannot capture everything when it comes to historical understanding. Looking at a paradigm case, I will argue that it is much harder to shut out meaning at the individual level than it might seem at first glance. Even though there may be a strong motivation for a non-cognitive theory of meaning, there is no reason to think that such accounts cannot be treated as complementing cognitive approaches rather than competing with them. Be that as it may, such theoretical tolerance in and of itself invites the question of just where the terrain of the cognitive ends and that of the non-cognitive begins.

MEANING WITHOUT INDIVIDUALS

For all of the quotation of Geertz's evocative phrasing in his theoretical papers, it is hard to read his work as a whole without appreciating the deep sense of ambivalence that he brings to certain key issues. For all of his claims about the need to treat culture as a text in need of reading(s), much of the Geertzian corpus is littered with appeals to the methodologically much more mundane.

Consider the interpretation of the cockfight. Geertz urges that we understand it as involving matters of status[7] and as a "metasocial commentary"[8] about hierarchical rank. There is a:

> migration of the Balinese status hierarchy into the body of the cockfight. Psychologically an Aesopian representation of the ideal/demonic, rather narcissistic, male self, sociologically it is an equally Aesopian representation of the complex fields of tension set up by the controlled, muted, ceremonial, but for all that deeply felt, interaction of those selves in the context of everyday life.[9]

Geertz proposes that this interpretation be taken as an instance of a textual analysis:

> In the case at hand, to treat the cockfight as a text is to bring out a feature (in my opinion, the central feature of it) . . . its use of emotions for cognitive ends.[10]

And, treating it as such, Geertz warns of the consequences of eschewing a "social mechanics" in favor of a "social semantics":[11]

> The culture of a people is an ensemble of texts, themselves ensembles, which the anthropologist strains to read over the shoulders of those to whom they properly belong. There are enormous difficulties in such an enterprise, methodological pitfalls . . . .[12]

All of this would seem concordant with Geertz's theoretical position that cultural analysis is closer to literary criticism than science as standardly understood, with a consequent openness of interpretation. Yet these famous quotations notwithstanding, the structure of "Notes on the Balinese Cockfight" belies a much more traditional (empirical) approach when it comes to evaluating the proposed interpretation.[13] For, in support of the thesis that "the cockfight . . . is fundamentally a dramatization of status concern . . . ,"[14] *seventeen* "facts"[15] are presented for which it is claimed that the "concrete evidence, examples, statements, and numbers that could be brought to bear in support of them is both extensive and unmistakable."[16]

Beyond this tension between the hermeneutic and the empirical that pervades Geertz's own texts, it is easy to forget that agency lurks behind the two most often quoted Geerztian mantras: culture as a web of significance;[17] and ethnography as a venture in thick description.[18] After all, the "webs" are of our own making. It is "man" that is to be taken to be "an animal suspended in webs of significance *he himself has spun*."[19] Of course, the standard way to sidestep the implication of agency in this quotation is to point to the notion of "thick description." Nowadays the phrase "thick description" is more often associated with Geertz than with Ryle, and yet it is hard to see how it can be of much help in explicating Geertz's approach. For Ryle, thick description involves probing below the surface level of behavior. As Ryle puts it (discussing Rodin's *le Penseur*):

> Very likely *le Penseur* was just not murmuring something under his breath or saying it in his head. But the question is, "What is the thick description of what he was essaying or intending in murmuring those words to himself?" The thin description "murmuring syllables under his breath," though true, is the thinnest possible description of what he

> was engaged in. The important question is "But what is the correct and thickest possible description of what *le Penseur* was trying for in murmuring those syllables?"[20]

I won't try to say what this view comes down to or whether it is compatible with Ryle's other views on the nature of mind.[21] Still, all the talk of "essaying," "intending" and "trying for" is talk on the individual level. *Geertz's* construal of Ryle's example stresses its parasitism on culturally meaningful categories through which behavior is "produced, perceived and interpreted, and without which they would not . . . in fact exist . . . ."[22] This move avoids intentionalism by grounding thick description on the notion of meaningful custom. But, in doing so, it leaves open the question of what account is to be given of meaningful custom itself.

Both Geertz and his defenders try to solve this problem with appeals to Wittgensteinian or Winchian considerations.[23] Geertz himself writes that:

> cultural forms find articulation . . . in various sorts of artifacts, and various states of consciousness; but these draw their meaning from the role they play (Wittgenstein would say their "use") in an ongoing pattern of life, not any intrinsic relationship they bear to one another.[24]

Of course, that leaves "use" hanging unanalyzed, and Geertz's Wittgensteinian leanings embrace the idea that such an approach to meaning has to connect use to rules.[25]

The appeal to rules is often bolstered by examples like the playing of chess. The identity of chess pieces, as well as their movement, so the story goes, only makes sense in the context of the rules of chess, and it is these rules that are constitutive of the game. As a consequence, intentions play no role in making the saying of "Checkmate" "mean" what it means in the context of a game of chess. So, too, with hand signals on the road: what makes an extended arm a signal are the rules of the road that constitute an extended arm as such a signal, rather than the driver's intention. Moreover, what makes this account appealing is a rendering that allows unintentional signaling (and checkmating) to count as cases of signaling and checkmating.

All of this is familiar territory, but when one pushes for the details of the relationship between meaning and rules, things quickly get murky. Cases of chess and road signals are misleading examples because they depend on explicit sets of rules. But, of course, *explicit* rules are few and far between. To work, the account also has to make an appeal to *implicit* rules.[26] Now whatever the other problems with the adequacy of these approaches as a way of capturing meaning (see the previous note), a major difficulty that they pose for a non-cognitivist theory is that they are simply too powerful. For, albeit inadvertently, such accounts sanction the wild proliferation of meaning without restriction. The problem is this, in brief: an account based on implicit rules has to be able to distinguish between features of the culture that are *constituted,* or at least *governed,* by such rules, as opposed to those features of the culture that merely *fit* such rules—otherwise the account of rules ends up as far too promiscuous. The "webs of significance"[27] end up as a feature of *all* aspects of life and we end up with the wild extravagance of the claim that "Everything is a sign, a luxuriant sprouting of signs; trees, clouds, faces, coffee-mills . . . are enameled with layers of interpretation which twist and knead the semantic dough."[28]

On such a view, meaning becomes an *ineliminable* feature of all aspects of how humans are to be understood.[29] The expansive reach of this claim is far from benign, for seeing meaning everywhere deprives us of the ability to consider theories of meaning that vary in their claims about the role of meaning itself. One of the problems with this view is that the profligacy of meaning excludes any role for *habit*.[30] And making room for habit is at the extreme of a continuum of problems which arises from treating meaning as knowable *a priori*.[31]

Now one way to sidestep all of these problems of tying down the "thick" is to push hard on a distinction between the "actual" and the "as if." Explaining by treating a system as if it were a system of a particular kind, without worrying whether it is in fact, is a successful explanatory strategy in much of science—a well worn route to insight is to ignore real mechanisms at work in a situation and rather abstract features of the situation by treating it as conforming to a model—a model that has features of regularity and tractability that may be lacking at the nitty-gritty level of actual mechanisms. Of course, to say that this gives insight is to allow that insight comes in different varieties, only one of which involves an accounting of the way mechanisms actually work.[32]

If we are only interested in an account "as if," then the considerations raised so far do not constitute telling objections. Instead, the issue becomes whether or not an account "as if" does enough for us. To aspire to an account "as if" is to aspire to a theoretical account unconstrained by considerations of realism, and within the Geertzian tradition there is a well known line of thought that looks very much like this. Meaning on this view is not something found in the culture but a theoretical category the anthropologist brings to the work of systematizing what he or she finds in the culture. (I don't claim this to be an exclusive line of reasoning in Geertz—it coexists with a "realist" stream of thought that locates meaning in the culture in a straightforward way.)

What price do we pay for this view?

One complaint about the Geertzian approach, raised by the anthropologist Roger Keesing, is that it forms the basis for a "temptation to reify a 'cultural system' into a causal force."[33] But, to the contrary, I think the problem is the exact opposite. Setting aside the extravagance of talk about cultural *systems*, the problem is how to treat meaning as any sort of theoretical entity outside the individual while preserving its causal force.

However, nothing makes the problem of putting the Geertzian perspective to such use more salient than Geertz's endorsement of a view due to Sorokin. For Geertz, meaning *is* a matter of expression: acts "signify," have "import" and, through them, things are "getting said."[34] But cultural interpretation is *not* a matter of grasping the significations and imports. Instead it is a theoretical construction one step removed from interpretation within the culture.[35] And what underscores the (scientifically) constructed nature of these theoretical entities is the fact that in the "real" world cultural systems (of meaning) and social systems (of causal interaction) don't exist as distinct complexes. Rather, they are theoretical constructions out of reality. Following Sorokin, Geertz argues that:

> because these two types of integration are not identical, because the particular form one of them takes does not directly imply the form the other will take, there is an inherent incongruity and tension between the two and between both of them and a third element, the pattern of

> motivational integration within the individual which we usually call personality structure.[36]

These important views imply that the anthropologist's constructions should not be understood as even a counterpart for anything obtaining in the lived reality of the culture. They are theoretical constructions to be treated "as if" *rather than causally efficacious*.[37] The issue we end up with, then, is this: if we treat meaning as a theoretical entity to keep it out of the head without reducing it, can it play a causal role within the warp and woof of a culture?

In fact, except on epistemic grounds, there is in principle no reason to rule out the causal efficacy of theoretical entities (in contrast to laws) *as long as we do not treat them on an "as if" basis,* and once we allow for such an account of theoretical entities we can refine and localize those entities to make them as historically specific as we want. That is not to say that we cannot aspire to a non-local account of culture as an abstract construct that is (as it were) a theory of everything. But such aspirations are not inconsistent with a view of meaning understood as a theoretical entity that is much closer to the ground.

But where are these entities to be located in the culture if they are not in people's heads *and* are not analyzed on an "as if" basis? Haven't we come, full circle, back to the problem of where to locate meaning? Consider an analogy. Solubility is a property that salt has in virtue of which it dissolves under suitably qualified conditions, where "suitably qualified" is spelled out in terms of a specified range of those conditions, and salt crystals have this property whether or not these conditions are satisfied. Indeed, the salt crystals before me have this property even if those conditions *never* obtain for them. Notwithstanding what is, solubility is a property that obtains counterfactually. (Never mind if this salt *is* never put in water, if it were to be in water . . . .) But if we think of meaning as a theoretical entity along this dispositional analogy, what bears the disposition? I will try to give an answer by way of another analogy, drawn from aesthetics.[38] Suppose you reject a theory of authorial intention in an account of interpretation in art. The artist may have a private meaning, but once the piece of art enters the public domain he or she no longer controls its (public) meaning. As Wimsatt and Beardsley put it:

> The poem is not the critic's own and not the author's (it is detached from the author at birth and goes about the world beyond his power to intend about it or control it). The poem belongs to the public. It is embodied in language . . . an object of public knowledge.[39]

The issue here is easily misinterpreted as a matter of epistemology: that we are ignorant of authorial intentions. But what is really at issue is the independent status of the poem—one with a "life of its own."

It is this conception of independence that makes plausible the limitations on the role of authorial intention that Wimsatt and Beardsley claim for their position, and with it the rejection of an expressive account of meaning in art and literature becomes plausible. For Wimsatt and Beardsley the poem has a status not unlike custom—and hence the relevance of the position here. A particular custom *might* have an origin based in something akin to authorial intention but, even if it did and we knew of it, here, too, the implications would be limited. Custom, once established, does not remain under the control of such "authors" even if they ever existed.[40] The common thread here for objects of interpretation, then, is this: whatever the private meanings attached to them, it is their status as public objects and consequent public meaning that is of interest.

But in what sense, then, does art have such a "meaning"? If we follow the dispositional view, meaning (be it one or many) is something that art has the capacity to produce. But what, and where? One approach is to think of art as something that has the capacity to cause associations in an audience. Of course, some audiences may have no associations, and others may have idiosyncratic associations. But we can specify what counts as an appropriate audience—a *suitably qualified* audience—even if such an audience is never actually instantiated. (That is to say, we can understand this notion of a suitably qualified audience in counterfactual terms just as we did in the case of salt that is never put in water.)

Of course, the difference between salt and art (and, with it, culture) comes into play with the notion of "suitably qualified." But the difference is not as great as one might at first think. At first blush we think science must begin with fixed dispositional properties of interest and then "look" for the range of its applicability. But there is nothing to prevent us running things in the opposite direction. That is to say, we might begin with a range of conditions and look for the properties (if any) exemplified in that range. Although perhaps with different emphases, in both science and art, this process is dialectical—we move back and forth between the two directions of attack. In that process the space is created to make it possible to attack both unitary and universal accounts of both properties and meaning. In the case of the latter, this forum for the "contestation" of meaning is two-dimensional: both the content of associations and the satisfaction conditions for counting as suitably qualified are up for grabs in the context of the struggle for hegemony, along with the constraints of coherence imposed by existing structures within the culture.

On the view I have been defending, we think of culture and ritual as having meaning while cashing this in on the basis of its capacity to evoke meaning under the right circumstances. Why not deny meaning at the cultural level and simply speak of it as *evoking* meanings? The difference between such a view and the view outlined here is that only the latter allows for a conception of cultural meaning that is a feature of the capacity of the culture as opposed to an artifact of the audience in which the capacity is expressed. But is this a distinction that does any work? I think so, because it seems to me that the contestation of meaning is not a battle that is waged mind-by-mind but, rather, takes place at an entirely different "level" within the culture. Here I have tried to provide an account of how meaning at such a level can be causally efficacious even as we think of it as a theoretical entity. But can such an account do enough for us to be able to dispense with a cognitive account of meaning altogether? Even if we can show that a non-cognitive account of meaning can do a lot of work for us, in the end I think that we can show that no such account can make do entirely without considerations of the cognitive entering in, if only as part of a complementary analysis.

In this regard, I have no comprehensive account to offer. Instead, I want to suggest that any exclusively non-cognitive approach is going to have to rule out letting an account of meaning in the head get in through the back door. How might this happen? In what follows I will try to illustrate the problem. As we will see, one way meaning in the head can come back into the picture is by way of appeals to "reasonableness." Another is by way of questions of what I call the "reach" of meaning. The cases I will be discussing involve the notion of meaning in both landscapes and housing in the nineteenth century, concentrating on England and the United States. To some extent the choice of this literature is arbitrary on my part. Nonetheless it is rich with examples that illustrate the challenges I think are faced by any realistic model that aims to locate meaning outside the head.

LANDSCAPES

Speaking of symbolic landscapes, nobody has put the issue more clearly than Meinig.

> Our (American) conception of landscape is informed by (among others) two icons—the New England Village and the Main Street of the Mid-West. Particular groupings of particular kinds of buildings *say* "New England" or "Main Street"—and carry with them whole sets of associations. In the case of the village, white houses around a church with steeple form the set. In the case of the street, the set is linear and red brick with store fronts and professional businesses set above. Of course *the* New England Village *qua* icon and Main Street *qua* icon had their origins in actual towns from which they spread. The spread was both geographical and cultural. New England Villages populate the Mid-West as well as New England and Midwestern Main Streets can be found as far west as California. But they also spread onto cards, pictures, calendars, and of course into advertising.[41]

Meinig poses two crucial questions about the process of transformation from the actual to icon. These are: "What are the means of selection of particular kinds of localities for idealizations?"[42] and "How do these chosen scenes become generalized . . . ?"[43] More generally, we can ask: what are the mechanisms by which meaning is first constructed; and, what are the mechanisms by which it is then disseminated? Meinig offers tentative programmatic suggestions for how to provide answers to these questions. He is much clearer about the issue of dissemination than the issue of choice, emphasizing the importance of examining literature as well as "popular" culture including magazines, advertising and cinema.[44] On the issue of selection he thinks one route may be by way of examining "the most important seats of the most influential vehicles of propaganda . . . ."[45]

However, while such an examination *might* provide an answer that explains how, for example, the New England Village was chosen over its alternatives, nonetheless it sidesteps the issue of just how the connection was made between the New England Village and its iconographic properties. To answer these kinds of questions, I suspect there will be no general theory. Instead, only the nitty-gritty of history will provide an answer to these kinds of questions for the New England Village and other cases examined case by case. But can we find a general answer when it comes to the issue of conditions of adequacy that range over such cases?

HOMES AND GARDENS

Amos Rapoport urges us to conceive of "the built environment as encoding cognitive schemata which need to be decoded in order to produce appropriate and congruent schemata in the minds of users . . . ."[46] Such a conception carries with it the idea that the built environment functions "as a system of non-verbal communication,"[47] and sometimes it undeniably plays just this role. Thus, Heather Tomlinson details how intentional architectural design resulted in:

> The use of castellated facades in the design of the Victorian prison . . . [as] a conscious attempt to make prisons appear repulsive, symbolizing on the outside the terrors to be expected within. In this way it was hoped that potential wrongdoers would be deterred.[48]

But no general theory of meaning in buildings and landscapes can be grounded in this way—not all buildings and certainly not all landscapes have a meaning in virtue of intentional encoding or even any encoding at all. Nor do they have meaning in virtue of communicative interests or functions, and yet they (may) have meaning—*but under what circumstances?*

Writing about the parlor in working-class homes in Sweden, Jonas Frykman and Orvar Löfgren argue that it was:

> not a pathetic attempt to imitate bourgeois life-styles;

> instead, the room had its own symbolic meaning in working-class culture. It was a cultural space separated from the drudgeries of everyday life, and to enter it meant being ritually transformed.[49]

And, writing about the privatization of the house in Victorian Britain, John Gillis asserts that "The threshold was now clearly marked by the whitened step and the polished door . . . ."[50] Frykman and Löfgren make an explicit claim about meaning. Gillis's claim is implicit.

These are surely eminently reasonable claims. The testimony of a contemporary is cited by Frykman and Löfgren:

> We lived mainly in the kitchen. The [other] room my parents used as parlor was meant for show, and you had to be very sick to get to lie down there. . . . We all slept in the kitchen and the other room was kept clean.[51]

And, as Gillis points out, " the mid-century emphasis on boundaries and the symbolic use of yards and gardens constituted a new departure."[52] But the emphasis should be placed here on the word "reasonable." These arguments depend crucially on making an inference from the outside about the meaning of the parlor and the threshold. If a threshold's being "clearly marked" is to be understood as more than a matter of physical marking, it is only because it is hard to see what else it could be doing under the circumstances.

I don't mean to be heavy-handed in making this point—perhaps these inferences are justified in the absence of other ways of making the argument. We lack data on just what the symbolic meaning of the parlor was in working-class culture, just as we do about the threshold. At best, we can make a contextual inference about what would seem reasonable in the circumstances. Yet I am hesitant to follow along. Not only is any appeal to reasonableness a blunt instrument in distinguishing between what is and what may be thought of *as if* it is unless we can appeal to considerations of rationality, but moreover such considerations will also need an account of beliefs and wants as an adjunct, and with them, of course, will come an account that is back in the head again.[53] *But can we do without the appeal to reasonableness?* The only hope of doing so will be if we can provide enough detail of how meaning is constructed in particular historical circumstances. But, even if we can do this, can we do so without letting individuals back in the end? In what follows I will illustrate how trying to do so comes to grief when we press on the details of just how it is that meaning is disseminated.

DOING WITHOUT REASONABLENESS

In both *Building the Dream* and *Moralism and the Model Home,* Gwendolyn Wright has provided a detailed social and cultural history of American domestic architecture, concentrating on the nineteenth century.[54]

Beginning in about 1840, domestic, middle-class architecture evolved from a distinct Puritan style into what we have come to know as the Victorian style. Victorian style itself was superseded in a slow transition to the modern home from the 1870s to the second decade of the twentieth century.

Wright's work is of interest here because it is an explicit attempt to provide an account of the meanings of architectural forms associated with these periods, as well as how these meanings came about and won out over competing meanings for the same architectural forms. For our purposes, it is also the best exemplar of the challenge of trying to pursue such a program without relying on any inferences from the "reasonable."

In New England colonial towns, a two-room house with loft bedrooms was common, with a chimney dominating the interior space.[55] For defensive reasons, the doors were strong and the windows small and sparse:

> Nature was not considered a gentle, inspiring force to be courted. . . . The wall of the house was decidedly a barrier to the outside; there was no thought of continuity between the interior and exterior domain.[56]

The interior of such homes was organized in accordance with Puritan anti-individualist principles. "Surveillance of one another was necessary."[57] As such, nearly all activity was concentrated in just one of the two downstairs rooms:

> The hall, or keeping room, was the center of the family's waking life: the place for cooking, eating, making soap and candles, spinning yarn and weaving homespun cloth,

> sewing shoes, repairing tools, keeping accounts, and reading Scripture. Women and men, children and servants, worked together, under each other's watchful eye.[58]

The change from this kind of housing to modern housing via what we have come to know as Victorian style can be characterized in a number of different ways.[59] In the period from about 1840 to 1910, changes in middle-class housing coincided with changes in work patterns, home production, the organization of cities, as well as notions of both women and nature.

In this process the construction of suburbs constituted a major event in the construction of culture, and while a transformation took place in the conception of nature from threat to benevolence, the openness and barrenness of grid development forced housing itself to play nature's role. The resulting focus on the house as the locus of cultural meaning coincided with a diminution in both work in general *in* the home, as well as in the production of goods *for* the home. So, while men increasingly worked away from the home, middle-class women were more involved in the consumption (and contemplation)[60] of products for use in the home rather than in their production. If the suburban home was the primary locus in the construction of a culture, women (isolated) in such homes were a primary agent. But they were also central subjects as well. For the transition from pre-Victorian to post-Victorian housing also involved a transition in the role and conception of women as well.

NATURE AND THE HOME

Writing about English society, Lenore Davidoff, Jean L'Esperance and Howard Newby correctly remark of the construction of the rural idyll and its counterposition to the "urban":

> It is a tribute to the endurance of this convention that even today, to many of us "rural" has pleasant, reassuring connotations—beauty, order, simplicity, rest, grassroots democracy, peacefulness, *Gemeinschaft*. "Urban" spells the opposite—ugliness, disorder, confusion, fatigue, compulsion, strife, *Gesellschaft*.[61]

Of course, these associations are not unqualified with regard to all of nature. Davidoff and her colleagues are writing of the English countryside. Constructions in the opposite direction obtain as well—as, for example, in the case of Appalachia. Here we have enduring associations of a very different kind, and Allen Batteau has shown how these, too, derive from a nineteenth-century literary construction that permeated the culture.[62]

Still, the development of our conception of Appalachia notwithstanding, in the nineteenth century, in America too, nature became redefined in a counterpoint to the "urban." At the same time, the development of manufacturing techniques, as well as distribution, made it possible for basic middle-class housing to be built with extravagant embellishments. Ornate finishings, both inside and outside the house, could be ordered and added on to the basic structure. Ironically, economies of scale allowed for uniformities in production techniques which made it possible to produce variation in housing.[63] Variation allowed for irregularity in housing to be realized and irregularity became a metaphor for nature.[64] Home, then, was more than the physical locus of retreat from the world of capital and production. It became part of the dominant metaphor of nature in contrast with that world:

> the natural home was to have the appearance of having grown up out of the earth's rugged and varied materials. Materials for the facade were chosen and put together in a way that was, in theory, imitative of nature's complex juxtaposition of color and texture.[65]

The Chicago builder George Garnsey urged the introduction of "breaks, jogs and angles, the more the better . . . ."[66] But, if such irregularity was a prescription to evoke nature, it also arose for a different reason as well. Victorian houses became increasingly to be seen as revealing the "personality" of their occupants—both by way of its exterior and, especially, in its interior through the parlor:

> Each bay, window, porch and other protrusion was considered evidence of some particular activity taking place within. . . . As the number of rooms in a moderate-cost suburban house increased, floor plans burst into extraordinary shapes.[67]

As the architects Palliser and Palliser reminded their readers, "Speaking of Home, what tender associations and infinite meanings cluster around that blessed word."[68] The problem was that

"Middle-class Victorians wanted to believe that their houses were unique. At the same time, certain patterns were necessary so that other people could clearly read the symbolism of social status and family life in the details."[69]

The social machinery by which these "patterns" were set and disseminated depended primarily on three popular institutions: the pattern book, the magazine and the domestic guide,[70] and the department-store display rooms. Of these, the clearest line of influence can be found by way of the pattern book.

In the middle of the nineteenth century, builders (often designating themselves as architects) began to publish books of house construction plans with accompanying commentary "that explained to readers the meaning of the form"[71]—much of it influenced by the associations prescribed by John Ruskin between architecture and the "domestic virtues." Thus, residential design was meant to emphasize:

> prominent chimneys and fireplaces, wide overhanging roofs, and bay windows—which . . . evoked the revered domestic virtues of protection, security, trust and traditional family bonds.[72]

Many of these elements had little functional purpose. For example, the popularity of the Victorian fireplace post-dated the introduction of furnaces and room stoves.[73] (The Berkshire Apartments built in New York in 1883 featured them in every room except the kitchens, even though the building was centrally heated.)[74]

The ideology of pattern book writers was that "The value of an object or view depended on deeply rooted symbolic associations it evoked."[75] And this associative capacity was taken as universally rooted,[76] rather than an artifact of their own construction. But, of course, in defending this universality, pattern book writers themselves provided an effective mechanism for the construction and dissemination of prescriptive interpretations, especially in view of the popularity of the medium at all levels of society.[77]

NATURE REDEFINED

If the transition to the Victorian style was gradual, so was the exit from it. Many of the defining themes persisted in the modern conception of the twentieth-century house. Less ornate and smaller, the notion of a balance between individuality and uniformity was none the less retained by providing for variations on basic designs.[78]

Considerations of economy loom large in driving many of these changes.[79] But other factors also obtained:

> In 1896, George E. Waring, Jr., who had recently been appointed head of New York City's Department of Street Cleaning, led his 2,700-man force, dressed in their new white uniforms and pushing their white garbage cans, down Broadway. Crowds cheered the parade of "Waring's White Wings" from the sidewalks.[80]

Such hoopla was part of a changed view toward nature as less benevolent in contrast with the Victorian era. If nature was seen as threatening to the pre-Victorians, it was seen so at a macroscopic level. But now the threat obtained at the microscopic level, in the form of the germ theory of disease.

This attitude not only affected public space but the home as well, and viewed through this gestalt, the Victorian house and, especially, features of its interior came to be seen as sources of danger. Bric-à-brac, ornate moldings and porous surfaces were hard to keep clean and thus to keep germ free.[81] "An architecture of visible health emerged in many domestic guides and home magazines. . . . The clearest symbol of cleanliness was the color white" writes Wright.[82]

The transformation of the "celebration of softness and abundance" into the ideal of "purity and simplicity"[83] also gained impetus from a coincidence among competing views of women. On one view, the traditional Victorian ideal of women as a force for the improvement of society was modernized by way of usurping scientific metaphors and applying them to the home.[84] The home, and especially the kitchen as laboratory, could help women better to realize their duties as they were traditionally conceived.[85] Yet at the same time:

> many different groups were campaigning for what they called a progressive approach to house design and upkeep. While their social goals were often based on conflicting values, public-health nurses, arts and crafts advocates, feminists, domestic scientists, and settlement-house

> workers favored the same simplified, standardized home to represent those values.[86]

Since "the home" was a central metaphor of the period,[87] what was at stake in these campaigns was more than just who would control what the "simplified, standardized home" would represent in the narrow sense of "home." The debate was about society at large and the role of women in it. This debate was prosecuted at a number of different levels. But like nineteenth-century discussions of home and house, it had its own machinery by which new "patterns" of housing and what they were to mean were set and disseminated. So while, as in the case of the Victorian style of house, the pattern book, the magazine and domestic guide were put to use,[88] they were now supplemented by both advertising and mail-order catalogues.[89]

THE QUESTION OF INDIVIDUALS

Gwendolyn Wright writes of "something peculiarly American about the campaign to define and publicize models for the home and to connect those models to other, larger social goals."[90] Peculiar or not, it provides a rich illustration of the machinery of meaning in action, with enough detail to obviate the need to appeal to considerations of reasonableness as a way of filling in the missing gaps. But does Wright's account of American architecture provide support for a view of meaning that is not located in the head? In what follows, I want to argue that, even if it does, it does so only at the cost of avoiding questions about meaning that can only be answered by bringing back individuals again.

One of the great virtues of Wright's history is the amount of attention she pays to the mechanisms by which meaning was both set and disseminated.[91] Contrast this to attempts that show changing conceptions of nature as reflected in literature, without attention to the role that this literature played in *setting* the conception *or* its dissemination.[92] Contrast it also to hermeneutic approaches to the city that provide no method to allow for their integration into the domain of the politics of the city.[93]

Still, there is one area in which this account may be seen to be less than complete. Dissemination can occur in one of two ways. One way is by the success of a particular interpretation over others. The other is by the success of a particular interpreter over others. The first countenances no division of labor. Everyone interprets, and which interpretations come to dominate is the issue. In the second, which interpretation comes to dominate follows which interpreter comes to dominate. Here, a division of labor *can* be tolerated. Not everyone need embrace an interpretation explicitly; they need only embrace the authority of the interpreter.

Is there any evidence of this distinction about dissemination in Wright's account? Let us assume for the purposes of argument that the Victorian-era views presented did in fact achieve hegemony (leaving aside the post-Victorian phase of "contested" meanings). But what was the reach of this meaning? Wright assures us of the popularity of pattern books and other mediums of dissemination.[94] Yet this popularity and consequent influence is compatible with two different stances toward the consumer at large. These are:

1 The promulgated interpretations were embraced by the population at large.
2 The promulgators of the interpretations were embraced by the population at large.

In the first stance the proposed interpretations were taken to heart by the population. They learnt the interpretations from the sources of dissemination and came to see things as they were "meant" to. In the second, they came to defer to the authority of those proposing the interpretations without necessarily coming to see things their way.

These, then, are the two mechanisms of dissemination. Unfortunately, there is also a third alternative that we must distinguish from both the first and the second. This is:

3 Neither (1) nor (2) took place.

This third stance is the view that hegemony *without influence* is possible. It is the view that those in control of the public culture could promulgate prescribed interpretations without such views or their own status being taken to heart. How is this possible? In what sense can hegemony be said to obtain under such circumstances? The third stance is an instance of the case in which the population at large attaches *no* meaning to features of the architectural world. Hegemony can obtain in matters of interpretation because of the absence of competition at the level of public meaning, where the reach of that hegemony is limited by the fact that not everyone in the society *is* engaged in spinning "webs of significance"[95] or in deferring to those that are.

To distinguish between these three alternatives, we need more than we get in Wright. There may seem to be an answer in the claim that: "All over the country, people of every class, from the mechanic to the dowager, had become familiar with the aesthetic theories of John Ruskin."[96] Yet I wonder whether it would not be more accurate to say that they had the opportunity to become so familiar? But did they avail themselves of it? And, even if they did, how did it effect the way they saw their buildings?

One way to try to find out would be by way of an account from the bottom up that allows us to examine the lived reality of daily life. Yet to do so is to pursue a route fraught with the danger of ensnarement in the idiosyncrasies of the individual. Still, if our interest is in a full account of how meaning works, we have to bring individuals into the picture—if we are to assess meaning's reach.

6

CONCLUDING WORRIES

Joyce Appleby, Lynn Hunt and Margaret Jacob have recently argued that we labor in the shadow of a tradition (that they call the "heroic model") that misleadingly thinks of truth as "absolute truth," of truth not just as "true enough for the time being" but "true always and absolutely."[1] Beyond this conception of truth, the heroic model carries a view about knowledge: namely that "things can be known in ways that correspond to actual objective existence" and that there is "a tight fit between nature and human knowledge."[2]

The counterpoint to this tradition, born, Appleby *et al.* argue, in reaction to it, is a family of relativisms that reaches its most extreme manifestation in the non-referential views about language championed by post-structuralists and associated views that undermine the value of truth-seeking as an enterprise understood as anything that has much to do with "truth."

In contrast to these polar alternatives, Appleby *et al.* offer a "third way," dubbed practical realism, which owes its core to a conception of pragmatism: truth emerges in a collective enterprise even if tentatively and contingently:

> people's perceptions of the world have some correspondence with that world and . . . standards, even though they are historical products, can be made to discriminate between valid and invalid assertions.[3]

It is no criticism of these authors to point out that, in recent years, there has been a number of attempts to pursue this line of argument. Kloppenberg criticized Novick for his failure to pay serious attention to pragmatism as a "third way" in *That Noble Dream*.[4] And even where pragmatism is not offered as the way out, there are striking similarities in the way in which the alternatives it is designed to replace are drawn in the literature.[5]

As I see it, the philosophical difficulty in this kind of approach is this: if you rely on pragmatism to ground an attitude toward inquiry that is appropriately tentative in nature, you thereby take an epistemological stance but you leave unsettled the issue of ontology unless you can argue that your epistemological position somehow rules out some ontological positions. Yet objectivism is an ontological doctrine and one that is consistent with at least one stream of pragmatism. Indeed, Peirce sounds like an advocate of full strength objectivism when he writes that:

> in all its progress, science vaguely feels that it is only learning a lesson. The value of Facts to it, lies only in this, that they belong to Nature; and Nature is something great, and beautiful, and sacred, and eternal, and real,—the object of its worship and its aspiration.[6]

While taken as a whole, American pragmatism is far too loose and contradictory a set of doctrines to be able to support a comprehensive alternative to either objectivism or relativism.[7] Nor is it clear that pragmatism offers a particularly unique route to the methodological prescriptions historians have been inclined to hang on it. Rejecting Peirce's logic of inquiry, Kloppenberg writes that:

> As James understood and Dewey reaffirmed, there is in pragmatic theory a fruitful alternative to relativism [and objectivity]. Hypotheses—such as historical interpretations—can be checked against all the available evidence and subjected to the most rigorous critical tests the community of historians can devise. If they are verified provisionally, they stand. If they are disproved, new interpretations must be advanced and subjected to similar testing. The process is imperfect but not random; the results are always tentative but not worthless. It is this strand of pragmatic hermeneutics, which has been present in the best work of American historians since the first decade of the twentieth century. . . . When all historians are categorized as either objectivists or relativists, this intermediate position may either disappear from view or seem a tepid compromise.[8]

Yet there are plenty of non-pragmatic ways to support the same view. For the view is about as mainstream as you can get in matters of methodology. You find it in both the traditions of Machean positivism and early twentieth-century versions of empiricism.[9] That should not be surprising, for these Jamesean sentiments arose in an American context of anti-idealism that was mirrored by parallel movements in Europe.

In the end, even if one can appeal to the pragmatic tradition for selective support of some methodological prescriptions, it is not clear what weight the appeal itself carries. Better to examine the merits of the doctrines advocated on their own, and in this regard one would be hard pressed to take issue with Appleby *et al.* or Kloppenberg in their advocacy of the merits of open, democratic, many-voiced inquiry that is suitably modest in its epistemic claims. But will these prescriptions ground any kind of ontological position?

Now you might respond by arguing that Appleby *et al.* or Kloppenberg (and indeed most historians) never had an interest in anything more than the epistemological.[10] But that would be to sidestep what is at stake in both the objectivists' stand and the relativists' response. For their debate is whether there *is* a fact of the matter in matters historical, not just whether we can know it.

What, if anything, can be said directly about this issue has been the subject of this book. Like Appleby *et al.* and Kloppenberg, I have defended a middle ground between metaphysical realism and anti-realism, and, like them, what I have defended might be thought of as a version of pragmatism as well—not in the sense of a formal doctrine but rather as an attitude toward the role of metatheory that is driven by historical practice. I have not argued that all history instantiates the middle position that I have developed. Nor have I claimed that it is necessary to interpret any history as doing so. Rather, my argument has been conditional—*if* you take certain historical interests seriously, then, realism as a style of reasoning comes along for the ride. In what follows I will pose a set of interrelated philosophical objections to this program and attempt to respond.

Think of historical practice(s) as our data—where that data includes what historians do—including building theories. Think of philosophy of history as a theory about all of this data. (Of course historians do philosophy of history—so let's shift this activity from data to theory about data.) Then realism can be thought of as part of a theory about the data of historical practice that is supported by that practice. So far so good. But if we look to a parallel in science between theory and data, two objections arise, both of which attack the power of data:

> *The argument treats evidential data as capable of mediating between competing theoretical claims. In so doing it assumes that there is a clear distinction that can be drawn between theory and observation. But there is no such neutral account of evidence to be had. What counts as data is constituted by theory and is not independent of it. As such, differing theories will have their own data. Each theory will have its own evidential "base" of support* (Objection 1). *Furthermore, even if there were a theory-neutral account of data to be had, the thesis of underdetermination of theory by data guarantees that there will always be at least some competitors for a favored theory that are equally well supported by the data. So, if a theory is underdetermined by data, then data cannot be relied on to distinguish between competing theoretical claims* (Objection 2).[11]

So it looks as though the main thesis is vulnerable to attack from two different directions. I introduced these objections by way of an analogy that treated practice in history and theories about history as parallel to data and theory about data in science. Is the analogy tight enough that the consequences of the former must be applied to the latter? A theory in science has to go beyond its instances to qualify as a theory. But does a theory about the practice of history have to have the same characteristics as a theory *in* science in order to count as a theory? Does it have to go beyond its instances? In fact we can ask the same question about a theory *of* scientific practice as opposed to a theory *in* science. Does it have to reach beyond its instances? A scientific theory can be thought of, at minimum, as a mechanism that allows projection from past data to future data. But why should we think that a theory *about* science, let alone history, ought to conform to this requirement? The reasoning depends on whether we approach a practice-driven view descriptively or prescriptively.[12]

Consider the descriptive case first. A descriptive theory of science or history that is practice-driven eschews appeal to any other considerations. Thus, *if* practice is variegated, a theory of science will have no obvious basis for whittling down this variation. The resulting "theory" will allow no projection from current practice to future practice except on the basis of probability—only past conformity will beget the expectation of future conformity. It is for this reason that I think that *purely* descriptive practice-driven models of science (or history) are not very interesting except in so far as practice displays some inherent homogeneity of its own (even if it does not do so explicitly).

But a prescriptive approach is quite a different matter. Here we restrict ourselves to practice as well and we do not allow theoretical considerations to enter in. Nonetheless, non-theoretical considerations are allowed to generate prescriptions that favor some practices over others. But these prescriptions are not just backward looking. They prescribe what ought to be in the future as well as what has been in the past. Hence they are projections even if they are not the standard projections of expectation, and, by parallel reasoning, the same considerations can be applied to theories about the practice of history.

So I think there is a way in which we can speak of philosophy of history standing to historical practice as a theory stands to data, and I think that the connection is close enough that we ought to worry about the force of the objections raised. The theory-ladenness of data (Objection 1) comes in two importantly different variants. One is a thesis about perception; the other is a thesis about language. It *may* be the case that we do not see in a theory-neutral way, so that those holding different theories will see different things; and it may be that we cannot separate our observation language from our theoretical language, so that the reference of observation terms will be different in the languages containing different theoretical terms. Let us assume both of these claims to be true for the purposes of argument, still, this would not be enough to create a problem unless we embrace these positions as totalistic. If *any* theoretical difference generated different perceptions, or *any* difference in theoretical terms affected the referent of observation terms, then no observational data could mediate between conflicting theoretical claims. And while such views have been asserted, if they were correct, our sense that data really makes a difference would turn out to be illusory.[13] But data certainly seems to make a difference. If we could be confident that we were not deluded about this, we could infer that the totalistic view about the relation between theory and data is wrong. But, of course, this is a thesis that is itself subject to what is at issue. So, answering this objection requires us to conditionalize on an anti-skeptical assumption: namely that data can in fact adjudicate between disagreement unless totalizing views about perception and reference can be answered directly (most likely by way of some sort of causal theory). But, whichever of these two routes one takes, it is important to emphasize that by rejecting a totalizing form of the argument we are not thereby automatically thrown into the arms of an atomistic position. Any form of holism short of its all-encompassing variant can allow for data to adjudicate between disagreement, as long as the arena of disagreement is not implicated in the theoretical background of the data,[14] and, while there is no guarantee that this should be the case, there is no *a priori* reason to assume that it is not.[15]

But, if competing theories may be viewed as not necessarily producing their own exclusive data, Objection 2 raises the worry about whether there can ever be enough data to make a difference. The thesis of the underdetermination of theory by data holds that, for any theory supported by any set of data, there will always be at least one competing theory that is both incompatible with the first theory and equally well supported by the same data. (It is easy to see how to instantiate the thesis in a trivial way: for any theory T supported by data d, construct a theory T^1, where the only difference between T and T^1 is that T^1 contains theoretical material that is incompatible with the theoretical material of T but isolated from d. The non-trivial variant is harder to construct, unless we allow that theories not only encompass all the actual data but also carry counterfactual force.[16] Then two theories might agree on all of the data while making differing assertions about what would have obtained under (for example) differing initial conditions.) Now the underdetermination theory may be broached as a global thesis or a local thesis. As a global thesis it is a thesis about a theory of everything. As a local thesis it is a thesis about anything less than its global variant. Let us assume that both variants hold without worrying too much about why this might be the case.[17] Then, how damaging will the thesis be? The only reason to care about the underdetermination thesis is if you are a realist. For absent realism, there is no discomfort in the conclusion that there may be no fact of the matter between competing theories. But I argued in Chapter One that realism is not a meaningful concept when it comes to high-level theories, and a global theory is the highest level of theory we can have—it is, as Quine puts it, a theory of everything. My argument was that, where realism matters is at less rarified levels—that is at the local level. But, *absent* a special condition that I will ignore here,[18] local underdetermination carries no anti-realist bite. Imagine a causal signal from a far-off galaxy (perhaps even a message from another civilization) that is directed toward us on Earth. Suppose it encounters a black hole en route. Causal signals don't escape from black holes. The information is lost. Our situation becomes epistemically underdetermined relative to the signal. We can never know what it said—whether, for example, it asserted *P* or not. But here our epistemic ignorance does not bleed into the ontological realm. There is no inclination to say that there was no fact of the matter. It is just that we cannot know the fact of the matter. That is just to say that epistemic underdetermination does not undercut ontological realism. Why should this be? I take it that the reason is that in situations like this we appeal to background knowledge that we import into the situation that has been accorded a realist status that is too robust to be undermined by epistemic quandaries. Likely views about causation lurk in the background here. Begin with a realist view about causation and all the foregoing will follow straightforwardly. Balk at the first move and the rest will be a problem. If you come to cosmology with background knowledge that grounds realism, epistemic underdetermination will not drive through to the ontological. And what holds for cosmology surely holds for history as well—for neither constitutes a theory of everything.

NOTES

INTRODUCTION

1 P. Novick, *That Noble Dream*, Cambridge, Cambridge University Press, 1988.
2 The subtitle of *That Noble Dream* is *The "Objectivity Question" and the American Historical Profession.*
3 *Ibid.*, pp. 1–2.
4 Here and throughout this chapter I treat "objectivity" and "realism" as synonymous.
5 However, Rorty is the first to acknowledge that he sees his views as an *extension* of a tradition that includes Quine, Davidson and Putnam. See R. Rorty, *Philosophy and the Mirror of Nature*, Princeton, Princeton University Press, 1979, p. 7.
6 See A. Fine, "And Not Anti-Realism Either," *Noûs*, 1984, vol. 18, pp. 51–66.
7 H. White, "The Fictions of Factual Representation," *Tropics of Discourse*, Baltimore, Johns Hopkins University Press, 1978, pp. 126–127.
8 Novick, *op. cit.*
9 *Ibid.*, pp. 25–31. However, this may not be true of the *very* first generation of American historians. (For more on this view see D. Ross, "On Misunderstanding of Ranke and the Origins of the Historical Profession in America," in G. Iggers and J. Powell (eds), *Leopold von Ranke and the Shaping of the Historical Discipline*, Syracuse, Syracuse University Press, 1990, pp. 154–169.) It is an irony that the "Rankean" dictum is now most often associated with an empiricist conception of history, given that Ranke himself was viewed in Germany as anything but an empiricist. (For more, see G. Iggers, *The German Conception of History*, Middletown, Wesleyan University Press, 1983, pp. 63–89.) Nothing brings out this opposition more than the question of how one translates the word "eigentlich" in the phrase "wie es eigentlich gewesen"—for, as Georg Iggers argues, one nineteenth-century sense of the term is "essentially." (G. Iggers, "Introduction" to L. Ranke, *The Theory and*

Practice of History, Indianapolis, Bobbs-Merrill, 1973, pp. xix–xx. See also Novick, *op. cit.*, p. 28.)

10 *Ibid.*, p. 31.

11 Even as it gives short shrift to other approaches to the discipline. (See L. Gordon, "Comments on *That Noble Dream*," *American Historical Review*, 1991, vol. 96, pp. 683–687.)

12 For now I use "fact" in the vernacular.

13 A good early treatment of their place can be found in C. Strout, *The Pragmatic Revolt in American History: Carl Becker and Charles Beard*, New Haven, Yale University Press, 1958.

14 Novick, *op. cit.*, p. 105. For Becker, see P. Snyder (ed.), *Detachment and the Writing of History, Essays and Letters of Carl Becker*, Ithaca, Cornell University Press, 1958. For Beard, see C. Beard, *The Nature of the Social Sciences*, New York, Scribner, 1934, pp. 50–72.

15 *Ibid.*, p. 167.

16 For a different view, see J. Higham, *History*, New York, Garland, 1985, p. 122.

17 Novick, *op. cit.*, pp. 254–255.

18 *Ibid.*, pp. 98 and 150–154. A more extended treatment is to be found in Strout, *op. cit.* I will come back to discuss the bearing of pragmatism on the argument in the last chapter.

19 Private communication.

20 C. Beard, "Written History as an Act of Faith," *American Historical Review*, 1934, vol. 39, p. 219.

21 C. Becker, "Everyman his Own Historian," *American Historical Review*, 1932, vol. 37, p. 233 (my emphasis).

1 OBJECTIVITY RECONFIGURED

1 P. Novick, *That Noble Dream*, Cambridge, Cambridge University Press, 1988, pp. 1–2.

2 However, like Megill (A. Megill, "Four Senses of Objectivity," *Annals of Scholarship*, 1991, vol. 8, pp. 301–320), you can define objectivity in a variety of ways, each of which has different costs and benefits associated with it. Nevertheless, I think that such an approach obscures rather than enlightens what is at issue. (For a wonderful treatment of the shifting notion of objectivity in the nineteenth and early twentieth centuries, related to the rise of mechanically produced images, see L. Daston and P. Galison, "The Image of Objectivity," *Representations*, 1982, vol. 40, pp. 81–128.)

3 For now I use "event" in the vernacular.

4 See, especially, M. Devitt, "Aberrations of the Realism Debate," *Philosophical Studies*, 1991, vol. 61, pp. 43–63.

5 *Pace* B. van Fraassen, *The Scientific Image*, Oxford, Clarendon Press, 1980.

6 *Pace* H. Putnam, *Reason, Truth and History*, New York, Cambridge University Press, 1981.

7 You can say that we "make" these truths. But even if we make them (in the sense that truths are statements and there exist no statements except in so far as we make them), when we make them, we make them as absolutely true only if they are in fact true.

8 J. Appleby, L. Hunt and M. Jacob, *Telling the Truth about History*, New York, Norton, 1994.

9 For an excellent discussion of why a so-called Tarskian construal of truth does not force any of these alternatives on us, see R. Kirkham, *Theories of Truth*, Cambridge, Bradford Books, 1995.

10 I do not mean here to be taken as implicitly defending an anti-holist view of language. The question is neutral between holist and atomist views of reference.

11 That is, the thesis that there is no fact of the matter between alternative translation schemas. See W. Quine, *Word and Object*, Cambridge, Mass., MIT Press, 1960.

12 In contrast to Murray Murphey. See M. Murphey, *Philosophical Foundations of Historical Knowledge*, Albany, SUNY Press, 1994.

13 See S. Kripke, *Naming and Necessity*, Oxford, Blackwell, 1980.

14 Thus, in baptismal variants, when parents name a child "Harry," they repeat the name to a clerk who inputs the data—but with an error, as "Hurry." The clerk's printer has a fault that causes it to print the letter "u" as "a." A causal chain is preserved, but it is not the right kind of causal chain because to come out right it depends on some accidental features.

15 M. Devitt, *Realism and Truth*, Princeton, Princeton University Press, 1984.

16 The best overall discussion can be found in H. Putnam, *Realism with a Human Face*, Cambridge, Mass., Harvard University Press, 1990.

17 *Ibid.*, pp. 118–119. In contrast, Lorenz (C. Lorenz, "Historical Knowledge and Historical Reality: A Plea for 'Internal Realism'," *History and Theory*, 1994, vol. 33, pp. 297–327) offers a misconceived answer to the challenge by illicitly treating internal realism as a variant of realism that "rests on basic presuppositions: first, that reality exists independently of our knowledge thereof; and second, that our scientific statements—including our theories—refer to this independently existing reality" (p. 308).

18 Including a reductive account of truth to an epistemic notion—which Putnam now views as a misinterpretation of his earlier positions. See H. Putnam, *Representation and Reality*, Cambridge, Bradford Books, 1988, p. 115.

19 However, certainly Putnam would disagree since his version of realism includes the claim that there is exactly one true and complete description of the way the word is with the concept of truth cashed in in terms of a correspondence. Call these two doctrines metaphysical realism$_2$ and metaphysical realism$_3$ respectively following Field (in H. Field, "Realism and Relativism," *Journal of Philosophy*, 1982, vol. 79, pp. 553–567). Then, following

Field, call metaphysical realism$_1$ the doctrine that the world consists of some fixed totality of mind-independent objects—essentially the doctrine with which we began less the correspondence account which we stripped away. Putnam (in H. Putnam, *Reason, Truth and History*, Cambridge, Cambridge University Press, p. 49) treats these three positions as part and parcel of one position. But, as Field argues, they need not be treated as such (*pace* Putnam in H. Putnam, *Realism with a Human Face*, Cambridge, Mass., Harvard University Press, 1990, pp. 30–32), nor have they been so treated—metaphysical realism$_2$ has little to recommend itself and metaphysical realism$_3$ has been denied by advocates of metaphysical realism$_1$ who embrace redundancy theories of truth.

20 A. Fine, "The Natural Ontological Attitude," in J. Leplin (ed.), *Scientific Realism*, Berkeley, University of California Press, 1984, pp. 83–107, "And Not Anti-Realism Either," *Noûs*, 1984, vol. 18, pp. 51–65, and "Unnatural Attitudes: Realist and Instrumentalist Attachments to Science," *Mind*, 1986, vol. 95, pp. 149–177. See also A. Fine, *The Shaky Game: Einstein, Reality and the Quantum Theory*, Chicago, Chicago University Press, 1986.

21 What follows is not an argument about the eliminability of the distinction between realism and anti-realism *ex post facto*. Hempel's theoretician's dilemma says you can always construct a version of a scientific account after the fact that undercuts support for a theory by providing a rendering of an equally well supported version of the account that is ontologically less grandiose. (See C. Hempel, "The Theoretician's Dilemma: A Study in the Logic of Theory Construction," *Aspects of Scientific Explanation*, New York, Free Press, 1965, pp. 173–226.) The theoretician's dilemma poses the challenge of how to develop a theory of justification to accommodate this fact. But here something else is going on—what is under discussion is the claim of eliminability *before the fact*.

22 Fine, "The Natural Ontological Attitude," pp. 95–96.

23 *Ibid.*, p. 101.

24 Fine, "Unnatural Attitudes," pp. 172–173.

25 *Ibid.*, p. 171.

26 A. Musgrave, "Noa's Ark—Fine for Realism," *The Philosophical Quarterly*, 1989, vol. 39, pp. 383–398.

27 R. Miller, "In Search of Einstein's Legacy: A Critical Notice of *The Shaky Game: Einstein, Reality and the Quantum Theory*," *Philosophical Review*, 1989, vol. 98, pp. 215–238.

28 *Ibid.*

29 Fine, *op. cit.*, pp. 172–173.

30 Miller, *op. cit.*, p. 235.

31 *Ibid.*, pp. 235–237.

32 Fine, *op. cit.*, pp. 172–173.

33 *Ibid.*

34 *Ibid.*

35 H. Stein, "Yes, but . . . Some Skeptical Remarks on Realism and Anti-Realism," *Dialectica*, 1989, vol. 43, pp. 47–65.
36 *Ibid.*, p. 50.
37 See *ibid.*, pp. 49–52.
38 *Ibid.*, p. 50.
39 *Ibid.* p. 58.
40 The phrase comes from Ian Hacking.
41 For more, see E. Sober, *The Nature of Explanation*, Cambridge, Bradford Books, 1984, pp. 47–51 and 71.
42 In conversation.
43 J. Appleby, "One Good Turn Deserves Another: Moving Beyond the Linguistic; A Response to David Harlan," *American Historical Review*, 1989, vol. 94, p. 1332.
44 J. Scott, "The Evidence of Experience," *Critical Inquiry*, 1991, vol. 17, p. 779. (The emphasis is mine.) I come back to these issues in Chapters Three and Five.
45 Why isn't this a variant of Boyd's strategy of defending realism by means of an inference to the best explanation of the success of science? (See R. Boyd, "The Current Status of Scientific Realism," *Erkenntnis*, 1983, vol. 19, pp. 45–90. In the same vein, see C. Lloyd, *The Structures of History*, Oxford, Blackwell, 1983.) That would be bad news—Boyd's strategy succumbs to the criticism that it begs the question—the inference only goes through if we embrace a realist view of explanation. In contrast to Boyd, I am not offering *explanation* of historical practice. Nor am I speaking of an inference to the external world as the explanans.
46 *Contra* Robert Berkhoffer (R. Berkhoffer Jr., *Beyond the Great Story*, Cambridge, Mass., Harvard University Press, 1995), who takes both of these features as the basis for an anti-realist approach.
47 Novick, *op. cit.*, pp. 599–607.
48 *Ibid.*, p. 523.
49 *Ibid.*, p. 544.
50 *Ibid.*, p. 546.
51 *Ibid.*, p. 554.

2 HISTORICAL FACTS

1 H. White, "Fictions of Factual Representation," *Tropics of Discourse*, Baltimore, Johns Hopkins University Press, 1978, p. 122.
2 A. Danto, *Narration and Knowledge*, New York, Columbia University Press, 1985, originally published as *Analytical Philosophy of History*. The book's importance in non-analytic traditions cannot be over-emphasized. It had a major influence on both Derrida and Habermas, and as we shall see, Danto's argument turns out to be much more powerful than other so-called "narrativist" approaches.
3 *Ibid.*, p. 140.
4 Of course, Danto is not alone in holding to the first incompatibility

thesis. We have already seen close ancestors of it in quotations from both Becker and Beard.

5 *Ibid.*, p. 142.
6 *Ibid.*, p. xiii.
7 *Ibid.*, p. 118.
8 *Ibid.*, p. 140.
9 *Ibid.*, p. 141.
10 *Ibid.*, pp. 132, 140, and 141.
11 *Ibid.*, p. 131.
12 *Ibid.*, p. 113.
13 Ibd., p. 146.
14 *Ibid.*, p. 155.
15 *Ibid.*, pp. 155–156.
16 *Ibid.*, p. 158.
17 *Ibid.*, p. 152.
18 *Ibid.*, p. 142.
19 *Ibid.*
20 N. Davis, "Women on Top," *Society and Culture in Early Modern France*, Cambridge, Polity Press, 1987, pp. 124–151. More generally, see F. Furet, *In the Workshop of History*, Chicago, University of Chicago Press, 1982, pp. 54–67.
21 This is perhaps because such accounts presuppose a familiar narrative in the background. See A. Megill, "Recounting the Past: 'Description,' Explanation and Narrative in Historiography," *American Historical Review*, 1989, vol. 94, p. 651.
22 *Ibid.*, pp. 644–645.
23 Unlike other narrative-based arguments.
24 As in the case of Danto, *op. cit.*, pp. 155–156.
25 *Ibid.*, p. 168.
26 *Ibid.*, p. 148.
27 A good place to start is M. Bradie, "Criteria for Event Identity," *Philosophical Research Archives*, 1983, vol. 9, pp. 29–78.
28 See D. Davidson, "Mental Events," in L. Foster and J. Swanson (eds), *Experience and Theory*, Amherst, University of Massachusetts Press, 1970, pp. 79–101.
29 See J. Kim, "Events as Property Exemplifications," in M. Brand and D. Walton (eds), *Action Theory*, Dordrecht, Reidel Publishing, 1976, pp. 159–177.
30 For more details see M. Brand, *The Nature of Causation*, Urbana, University of Illinois Press, 1976.
31 Danto, *op. cit.*, p. 155.
32 At least it is on the middle road that Danto chooses. There is the other middle road to consider—that in which events change but facts are permanent. But I don't take this to be a serious alternative.
33 Why not just say that history is about events and consists of facts about these events? That is to say, history is not about facts, but uses facts. So there is no sense in which the Past changes—at most, the significance we attach to events changes. I think this is wrong-

headed for two reasons. First, not all historical subjects of interest are captured by the idea of "an event." (Writing the history of the Renaissance *may* be to write the history of an event—but I certainly don't think the same can be said about the history of the English working class.) Second, events themselves seem to me to be eliminable from historical discourse altogether. Let a fact be a proposition that is true at a particular location or set of locations in space–time. (Even facts that are true of the English working class can be understood as true at . . .) Then the need for the notion of an event drops out of consideration. We can still have historical subjects (World War II), now understood as sets of propositions true at times and places. (And from here we can go on to be eliminativists about "truth" to boot.)

34 However, proceeding this way institutes a philosophically suspect distinction between "real" changes and so-called "Cambridge changes" that has proved notoriously hard to defend in a non-question-begging way.
35 At a more general level, see also M. Broszat and S. Friedländer, "A Controversy about the Historicization of National Socialism," *New German Critique*, 1988, vol. 44, pp. 85–126.
36 See K. Schleunes, *The Twisted Road to Auschwitz*, Urbana, University of Illinois Press, 1970.
37 For more details on this section see *ibid.*
38 *Ibid.*
39 A. Mayer, *Did the Heavens Not Darken?*, New York, Pantheon, 1988.
40 Am I merely playing Butterfield (H. Butterfield, *The Whig Interpretation of History*, New York, Scribner, 1951) to Danto's attenuated Whiggism here? I think not. For, unlike Butterfield, I hold no wholesale brief against the study of the past with reference to the present. (Of course, Danto is no wholesale Whig either—but the principle of the re-alignment of the Past depends on the principle of the study of the past with reference to later events, even if they are not necessarily events of the present.)
41 Nor is this the only slice to which this argument may be applied. Even in the more straightforward domain of the non-phenomenological, distance may give rise to comparative stability. Consider facts about the biological causes of the Black Death and their relationship to, at least some, later (temporally) local events. Of course, even here, much is still open to dispute, both as to some of the causes and some of the effects. For example, what was the role of urbanization as a cause, and what was the effect on the labor market? But this is not to say everything is still unsettled. Nor is it to say that what goes on now can in any interesting sense prompt a "retroactive re-alignment" of the relationship between these past local events—*even where some of the features of their relationship are as yet epistemically unsettled*—and that even applies if it should turn out that the cause itself was not *Yersinia pestis*. (For an alternative

account of the biology, see G. Twigg, *The Black Death: a biological reappraisal*, London, Batsford, 1984.)

3 THE CONSTRUCTION OF HISTORY

1 P. Novick, *That Noble Dream*, Cambridge, Cambridge University Press, 1988.
2 J. Scott, *Gender and the Politics of History*, New York, Columbia University Press, 1988.
3 This is in contrast to Scott's more recent work, *Only Paradoxes to Offer*, a book that is much more philosophically restrained than *Gender and the Politics of History* and yet remarkable in its intellectual ambition to demonstrate the way in which discursive practices and categories limit and shape the conceptual alternatives open to participants in a culture. (See J. Scott, *Only Paradoxes to Offer*, Cambridge, Mass., Harvard University Press, 1996.)
4 Scott, *Gender and the Politics of History*, p. 10.
5 *Ibid.*, p. 3.
6 *Ibid.*, p. 4.
7 *Ibid.*, p. 2.
8 *Ibid.*
9 *Ibid.*, p. 5.
10 *Ibid.*, p. 4.
11 *Ibid.*, p. 2.
12 *Ibid.*, p. 10 (my emphasis).
13 *Ibid.*, p. 7.
14 *Ibid.*, pp. 2–3.
15 L. Gordon, Review of *Gender and the Politics of History*, *Signs*, 1990, vol. 15, pp. 853–858. The quotation is from p. 854.
16 J. Scott, "Response to Gordon," *Signs*, 1990, vol. 15, pp. 859–869. The quotation is from p. 859.
17 See especially B. Palmer, *Descent into Discourse: The Reification of Language and the Writing of Social History*, Philadelphia, Temple, 1990.
18 Scott, *op. cit.*, p. 859.
19 And explicitly, in E. Albertson, D. Abraham and M. Murphy, "Interview with Joan Scott," *Radical History Review*, 1989, vol. 45, pp. 41–59. The quotation is from p. 50.
20 Originally published in *International Labor and Working Class History*, 1987, vol. 32, pp. 39–45, reprinted in *Gender and the Politics of History*, pp. 53–67.
21 G. Stedman Jones, *Languages of Class: Studies in English Working Class History, 1832–1982*, Cambridge, Cambridge University Press, 1983.
22 Scott, *Gender and the Politics of History*, p. 57.
23 *Ibid.*
24 *Ibid.*, p. 66 (my emphasis).

25 *Ibid.*
26 *Ibid.*, p. 5.
27 T. Kuhn, *The Structure of Scientific Revolutions*, Chicago, University of Chicago Press, 1962.
28 Although not by design. Kuhn does have an evolutionary tale to tell in an attempt to provide a basis for comparing theories, but I think it is fair to say that it is not entirely consistent with his core theory.
29 *Ibid.*, p. 150.
30 Although it may be a matter of historical dispute.
31 *Ibid.*, p. 151, quoting Max Planck.
32 See especially H. Putnam, "The Meaning of 'Meaning,'" *Mind, Language and Reality, Philosophical Papers*, Volume 2, Cambridge, Cambridge University Press, 1975, pp. 215–271. I think that the most promising idea offered so far is Dudley Shapere's "chain-of-reasoning-connection" approach. See D. Shapere, "Reason, Reference and the Quest for Knowledge," *Philosophy of Science*, 1982, vol. 49, pp. 1–23. (Of course, one significant difference between this argument applied to the subject matter of science and the argument as applied to history is that only the former might be thought to deal with natural kinds that have essential properties. But I do not think such views make a difference *here*. You do not need to assume natural kinds to be faced with the problem of how it is we can talk about differing views about the "same thing.")
33 For an example of this theme, see John Gillis's treatment of how the meaning of marriage has changed over time in British culture and the consequent danger of making temporally wide-ranging generalizations that assume the stability of meaning of terms like "bastardy." See J. Gillis, *For Better, For Worse*, New York, Oxford University Press, 1985.
34 Scott, *op. cit.*, p. 67.
35 Here I use the term "historicism" in what I take to be its current sense, that is, as no more than localism.
36 *Ibid.*, pp. 48–49.
37 Unless we eschew talk of kinds and instead allow that it is the *same* historical entity that is referred to by different tokenings of a term. (The idea comes from David Hull. See D. Hull, "Central Subjects and Historical Narratives," *History and Theory*, 1975, vol. 14, pp. 253–274.) But taking this alternative route will not avoid the problem under discussion—instead, it just shifts from a claim about reference to the same *kind* to a claim about reference to the same *entity*.
38 In Scott, *op. cit.*, pp. 93–112.
39 *Ibid.*, p. 96.
40 *Ibid.*, p. 103.
41 *Ibid.*, p. 100.
42 *Ibid.*, p. 101.
43 *Ibid.*, p. 102.

44 *Ibid.*, p. 107.
45 *Ibid.*
46 *Ibid.*, p. 108.
47 *Ibid.*, pp. 94–95 (my emphasis).
48 *Ibid.*, p. 7.
49 *Ibid.*, pp. 2–3.
50 *Ibid.*, pp. 94–95.
51 M. Poovey, "Feminism and Deconstruction," *Feminist Studies*, 1988, vol. 14, pp. 51–65.
52 *Ibid.*, p. 52.
53 *Ibid.*
54 *Ibid.*, p. 53.
55 *Ibid.*, p. 52.
56 *Ibid.*
57 Scott, *op. cit.*, p. 2.
58 Poovey, *op. cit.*, p. 58.
59 *Ibid.*
60 *Ibid.*, p. 62.

4 FOUCAULT BY HISTORIANS

1 J. Weeks, "Foucault for Historians," *History Workshop*, 1982, vol. 14, pp. 106–119; and J. Weeks, *Sex, Politics and Society*, London, Longman, 1989.
2 Weeks, "Foucault for Historians," p. 110.
3 *Ibid.*, p. 112.
4 *Ibid.*
5 *Ibid.*, p. 114.
6 *Ibid.*, p. 117.
7 *Ibid.*, p. 118.
8 Weeks, *Sex, Politics and Society*, p. 21.
9 *Ibid.*, p. 109.
10 However, Weeks himself reports "the emergence of a distinctive male homosexual subculture in some of the larger cities" in the 1720s (*ibid.*, p. 100). Weeks does not suggest how to reconcile this fact with the more general claim of the nineteenth-century construction of the homosexual as a person. The issue of when the emergence of the homosexual as a person occurs cross-cuts a number of more general conceptual issues that I do not take up here but that are superbly treated in M. Duberman, M. Vicinus and G. Chauncey Jr. (eds), *Hidden from History: reclaiming the gay and lesbian past*, New York, Signet, 1989.
11 Weeks, *op. cit.*, p. 98.
12 See *ibid.*, pp. 99–101.
13 *Ibid.*, p. 104.
14 *Ibid.*, p. 108.

15 *Ibid.*, quoting M. Foucault, *The History of Sexuality, Vol. 1, An Introduction*, London, Allen Lane, 1979, p. 101.
16 Weeks, *op. cit.*, p. 24.
17 *Ibid.*, p. 26.
18 *Ibid.*, p. 48.
19 *Ibid.*, p. 72.
20 *Ibid.*, p. 98.
21 *Ibid.*, p. 199.
22 *Ibid.*, p. 203.
23 *Ibid.*, p. 160.
24 M. Poovey, *Uneven Developments*, Chicago, University of Chicago Press, 1988.
25 *Ibid.*, pp. 8–9.
26 See *ibid.*, chapters 2 and 6.
27 I use this functional term deliberately for it is implicit in much of Poovey's approach, and explicit in places. See, for example, *ibid.*, pp. 50 and 171.
28 M. Poovey, "Feminism and Deconstruction," *Feminist Studies*, 1988, vol. 14, p. 52.
29 *Ibid.*
30 *Ibid.*, p. 10.
31 *Ibid.*, p. 17.
32 M. Foucault, *The Archaeology of Knowledge*, New York, Pantheon, 1972.
33 M. Foucault, *Discipline and Punish*, New York, Vintage, 1979.
34 I have used "determination" and "determinants" here to avoid stepping on Foucault's commitment against both the causal and the study of "origins." I will have more to say about this commitment later.
35 I use "anti-phenomenological" to cover a variety of sins associated with grounding talk about meaning, experience and point of view.
36 Of course, "presentism" can be interpreted in different ways, and here I am deliberately slurring them. Foucault's intended sense is much stronger than the idea that our presentist *interests* dictate our historical interests. But I think much of what makes him hold to a more contentious view is a product of his conception of knowledge—of which more below.
37 Foucault certainly did not.
38 J. Scott, *Gender and the Politics of History*, New York, Columbia University Press, 1988, p. 2.
39 *Ibid.*, p. 10.
40 *Ibid.*, p. 2.
41 Foucault, *The Archaeology of Knowledge*; see especially pp. 29, 46, 47, 90–91, 98–99, 108, 111, 164, and 200.
42 In M. Foucault, *Politics, Philosophy, Culture: Interviews and Other Writings of Michel Foucault*, New York, Routledge, Chapman and Hall, 1988.
43 *Ibid.*, p. 95.

44 Of course, even on the weak view, the result may be the same if the scope of your discourse falls within the subject matter of that discourse. Indeed, such a problem may arise with *The Order of Things*, even if its scope is restricted.
45 Poovey, *Uneven Developments*, p. 23.
46 In conversation.
47 J. Merquior, *Foucault*, London, Fontana, 1985.
48 H. Putnam, *Reason, Truth and History*, Cambridge, Cambridge University Press, 1981, p. 121.
49 A more attenuated argument to a less extreme negative conclusion about the status of *The Archaeology* is offered by Dreyfus and Rabinow in H. Dreyfus and P. Rabinow, *Michel Foucault, Beyond Structuralism and Hermeneutics*, Chicago, University of Chicago Press, 1983.
50 At least it can't given the high intellectual regard in which Foucault was held.
51 M. Foucault, "Nietzsche, Genealogy, History," reprinted in *The Foucault Reader*, New York, Pantheon, 1984, pp. 76–100.
52 *Ibid.*, p. 78.
53 R. Rorty, "Foucault and Epistemology," in D. Hoy (ed.), *Foucault, a Critical Reader*, Oxford, Basil Blackwell, 1986, p. 46.
54 A. Megill, *Prophets of Extremity*, Berkeley, University of California Press, 1985.
55 L. Shiner, "Reading Foucault: Anti-Method and the Genealogy of Power-Knowledge," *History and Theory*, 1982, vol. 21, pp. 382–398.
56 Foucault, *Discipline and Punish.*
57 Foucault, *The Archaeology of Knowledge*, p. 90.
58 *Ibid.*, p. 91.
59 See *ibid.*, pp. 98–99.
60 *Ibid.*, p. 224.
61 *Ibid.*, p. 218.
62 *Ibid.*, p. 108.
63 G. Gutting, *Michel Foucault's Archaeology of Scientific Reason*, Cambridge, Cambridge University Press, 1989.
64 *Ibid.*, p. 273.
65 *Ibid.*
66 See Foucault, *op. cit.*, p. 188.
67 *Ibid.*, p. 108.
68 *Ibid.*, p. 111.
69 *Ibid.*, p. 218.
70 *Ibid.*
71 *Ibid.*
72 *Ibid.*, p. 224.
73 *Ibid.*
74 *Ibid.*
75 *Ibid.*, p. 223.
76 Like me, Prado (C. Prado, *Starting with Foucault: an Introduction to Genealogy*, Boulder, Westview Press, 1995) is worried about how

Foucault can avoid the use of truth in an ordinary way in reading him on penal reform—despite the apparent inconsistency with his claims about truth itself (*ibid.*, p. 140) as well his use of "objectivist," "true" and "truth" (*ibid.*, p. 141). Prado's proposal is that we think of Foucault as not offering "an account of the nature of truth" at all (*ibid.*, p. 145). But then Prado goes on to endorse the view that Foucault's use of truth is linguistic and "commendatory," and that such a notion of truth can be put to use independent of genealogical interests and independent of "how things are true" (*ibid.*, p. 144). If Foucault is to be taken not to be offering a particular view of truth, why not leave it at that?

77 I have been concentrating here on *The Archaeology*; however, I do not think the same claims can be made about *The Order of Things*. Notwithstanding the latter's restriction in scope to the "human" sciences, whether by design or not, it contains a much more global argument about the nature of language than that which one finds in *The Archaeology*.

78 Foucault, *op. cit.*, p. 2.

79 *Ibid.*

80 *Ibid.*, p. 163.

81 And so, too, with experience itself. See *ibid.*, p. 47.

82 *Ibid.*, p. 117.

83 Foucault, *Discipline and Punish*, p. 104.

84 *Ibid.*, p. 112.

85 In fact, to the contrary, the assumption is usually that such considerations support a realist view—but we need not make that assumption here. Here, all we need to assume is that considerations of causation are neutral between realism and anti-realism.

86 "*that which is made*, the object, explains itself by the *making* of it at each moment in history; it is our mistake that we imagine that the *making*, the practice, explains itself by what is made" (P. Veyne, *Comment on écrit l'histoire, suivi de Foucault révolutionne l'histoire*, Paris, Éditions du Seuil, 1978, p. 219).

87 One place where the parallel *is* close is in the treatment of the Hegelian concept of history. For Derrida's contrast of this "metaphysical concept of history" and its alternatives, see J. Derrida, *Positions*, Chicago, University of Chicago Press, 1981, pp. 50–51 and pp. 56–58.

88 J. Culler, *On Deconstruction*, Ithaca, Cornell University Press, 1982.

89 J. Derrida, *Writing and Difference*, Chicago, University of Chicago Press, 1978, pp. 292–293. Quoted (in a slightly different translation), in Culler, *op. cit.*, pp. 131–132.

90 *Ibid.*, p. 132.

91 *Ibid.*, p. 133.

92 See also, for example, C. Norris, *The Content of Faculties*, London, Methuen, 1985, pp. 55–57.

93 In both the sense of the speaker as well as concepts.

94 See J. Ellis, *Against Deconstruction*, Princeton, Princeton University Press, 1989, pp. 60–62.

95 Although, as Graham Priest argues (in G. Priest, "Derrida and Self-Reference," *Australasian Journal of Philosophy*, 1994, vol. 72, pp. 103–111), the seeming unavoidability of the application of Derridean analysis to itself might be assimilated to more general problems of self-reference, prompting consideration of "standard" solutions like metalinguistic ascent and truth–value gaps. For a good survey of the problems associated with these "solutions," see again R. Kirkham, *Theories of Truth*, Cambridge, Bradford, 1985.

96 G. Prakash, "Can the 'Subaltern' Ride? A Reply to O'Hanlon and Washbrook," *Comparative Studies in Society and History*, 1982, vol. 34, pp. 168–184. The quoted passage is from p. 172.

97 *Ibid.*

98 Should the Subaltern Studies Group be understood as being influenced by deconstruction at all? After all, writing in the preface to volume one of *Subaltern Studies*, Ranajit Guha refers to Gramsci's six-point project in "Notes on Italian History." But I think his more significant claim is that: "We recognize of course that subordination cannot be understood except as one of the constitutive terms in a binary relationship of which the other is dominance . . . " (R. Guha, "Preface," *Subaltern Studies*, 1982, vol. 1, p. vii).

99 G. Spivak, "Subaltern Studies: Deconstructing Historiography," *Subaltern Studies*, 1985, vol. 4, p. 338.

100 *Ibid.*, p. 340. The quoted passage is from J. Derrida, *Of Grammatology*, Baltimore, Johns Hopkins University Press, 1976, p. 93.

101 Originally a collective made up of Shahid Amin, David Arnold, Partha Chatterjee, Ranajit Guha, David Hardiman and Gyan Pandey.

102 See *Subaltern Studies*, 1982, vol. 1, p. 3.

103 Prakash, *op. cit.*

104 Prakash suggests (in G. Prakash, "Subaltern Studies as Postcolonial Criticism," *American Historical Review*, 1994, vol. 99, pp. 1475–1490) that the interest is in recovering the subject and restoring agency, but that "the subalternist search for a humanist subject–agent frequently ended up with the discovery of the failure of subaltern agency: the moment of rebellion always contained the moment of failure. The desire to recover the subaltern's autonomy was repeatedly frustrated because subalternity, by definition, signified the impossibility of autonomy . . . " (*ibid.*, p. 1480). But notice how this runs together two different notions: autonomy and agency. Agency is the same as autonomy in only one of two senses. The other sense is associated with reason, and it is the second sense that is the important one—one related to consideration of reason quite independent of its success. (For a more general critique of Prakash's treatment of the relationship between agency and post-modernism, see R. O'Hanlon and D. Washbrook, "After Orientalism: Culture,

Criticism and Politics in the Third World," *Comparative Studies in Society and History*, 1992, vol. 34, pp. 141–167.)

105 S. Amin, "Approver's Testimony, Judicial Discourse: The Case of Chauri Chaura," *Subaltern Studies*, 1987, vol. 5, pp. 166–167.

106 In this regard, see especially R. Guha, "The Prose of Counter-Insurgency," *Subaltern Studies*, 1983, vol. 2, pp. 1–42; and "Dominance Without Hegemony and Its Historiography," *Subaltern Studies*, 1989, vol. 6, pp. 210–309.

107 A point the colonial powers never took seriously in the way they viewed the rebels. See, for example, D. Arnold, "Rebellious Hillmen: The Gudem-Rampa Risings 1839–1924," *Subaltern Studies*, 1982, vol. 1, pp. 140–141. (Nor, argues Gyanendra Pandey, did writers, be they colonial or nationalist. See G. Pandey, "The Prose of Otherness," *Subaltern Studies*, 1994, vol. 8, pp. 188–221.)

108 Even if its reconstruction can be neither full blooded in nature nor straightforward in method.

109 Arnold, *op. cit.*, pp. 88–89.

110 G. Pandey, " 'Encounters and Calamities': The History of a North Indian *Qasba* in the Nineteenth Century," *Subaltern Studies*, 1983, vol. 3, pp. 221–270; and G. Bhadra, "Four Rebels of Eighteen-Fifty-Seven," *Subaltern Studies*, 1985, vol. 4, pp. 229–275.

111 *Ibid.*, pp. 140–141.

112 D. Chakrabarty, "Trade Unions in a Hierarchical Culture: The Jute Workers of Calcutta, 1920–50," *Subaltern Studies*, 1984, vol. 3, pp. 116–152.

113 S. Henningham, "Quit India in Bihar and the Eastern United Provinces: The Dual Revolt," *Subaltern Studies*, 1983, vol. 2, p. 151.

114 G. Pandey, "Peasant Revolt and Indian Nationalism: The Peasant Movement in Awadh, 1919–1922," *Subaltern Studies*, 1982, vol. 1, p. 154.

115 G. Pandey, "Rallying Round the Cow: Sectarian Strife in the Bhojpuri Region c.1888–1917," *Subaltern Studies*, 1983, vol. 2, pp. 62–64.

116 *Ibid.*, pp. 85–86.

117 *Ibid.*, pp. 166–178.

118 R. Guha, "Forestry and Social Protest in British Kumaun, c.1893–1921," *Subaltern Studies*, 1985, vol. 4, pp. 93–100.

119 *Ibid.*, p. 100.

120 D. Hardiman, "The Bhils and Shahukars of Eastern Gujarat," *Subaltern Studies*, 1987, vol. 5, pp. 3–4, 22–23.

121 Guha, "The Prose of Counter-Insurgency," p. 2.

122 Spivak, *op. cit.*, pp. 338–339.

123 *Ibid.*, p. 338.

124 *Ibid.*

125 *Ibid.*, p. 340.

126 *Ibid.*, p. 341.

127 *Ibid.*, p. 342.

128 Guha, *op. cit.*, p. 2. However, see Rosalind O'Hanlon (R. O'Hanlon,

"Recovering the Subject Subaltern Studies and Histories of Resistance in Colonial South Asia," *Modern Asia Studies*, 1988, vol. 22, pp. 189–224) for reservations. O'Hanlon sees the Subaltern Studies series as unified in promoting the centrality of the task of reconstruction of the "subaltern as a conscious human subject–agent . . . " although as a project "in the form of the classic unitary self-constituting subject–agent of liberal humanism," its "limitations have yet to receive any proper public discussion" (*ibid.*, p. 196).

129 G. Spivak, "Can the Subaltern Speak?" in C. Nelson and L. Grossberg (eds), *Marxism and the Interpretation of Culture*, Urbana, University of Illinois Press, 1988, pp. 307–308.

5 EXTENDING THE ARGUMENT

1 In "The Evidence of Experience," Joan Scott takes issue with those "who have argued that an unproblematized 'experience' is the foundation of their practice . . . " (J. Scott, "The Evidence of Experience," *Critical Inquiry*, 1991, vol. 17, p. 780). For her, historical practice ought not to just be a matter of "making experience visible" but instead providing an "analysis of the workings of this system and of its historicity" (*ibid.*, p. 779). Moreover, Scott thinks pursuing an interest in the former precludes doing the latter (*ibid.*). *Prima facie*, this is a puzzling claim. Why should attempts to make experience visible *preclude* the study of its historical determination? Even if our interest *is* "to historicize it [experience] as well as to historicize the identities it produces" (*ibid.*, p. 780), it is hard to see how this project could get off the ground in the first place without the data of experience in hand. Even if Scott is correct that "experience" is not *evidentially* foundational, or situated in *autonomous* individuals, it is—*contra* Scott (*ibid.*, p. 779)—individuals who have experience. And if (as she suggests) our primary interest in experience ought to be the deconstruction of that experience and its underlying categories of construction, taking experience seriously looks like the plausible starting point for such a project. (Perhaps Scott means this: individuals are not the starting point. We experience as—men, women, and so on, and these are subject positions that are themselves the product of social discourse. Still, all of this notwithstanding, "preclude" seems far too strong a claim, while the anti-"individualism" comes down to this point: there are no individuals before experience and experience is a social product. Yet that does not prevent it being the case that individuals are the things that "have" the experiences—even if social construction turns out to have the reach that Scott thinks it has.)

2 *Contra* Collingwood's conception of history as exclusively constituted by the re-enactment of past thought. (See R. Collingwood, *The Idea of History*, London, Oxford University Press, 1946, p. 215.)

3 See M. Weber, *The Theory of Social and Economic Organization*, New York, Free Press, 1964, p. 89: " 'Meaning' may be of two kinds. The term may refer to the actual existing meaning in the given concrete case of a particular actor, or to the average or approximate meaning attributable to a given plurality of actors, or secondly to the theoretically conceived *pure type* of subjective meaning attributed to the hypothetical actor or actors in a given type of action."
4 The best analysis of Foucault in this respect is to be found in Veyne. See P. Veyne, *Comment on écrit l'histoire, suivi de Foucault révolutionne l'histoire*, Paris, Éditions du Seuil, 1978, p. 219.
5 J. Habermas, *On the Logic of the Social Sciences*, Cambridge, Mass., MIT Press, 1988, p. 87.
6 S. Ortner, "Theory in Anthropology since the Sixties," *Comparative Studies in Society and History*, 1984, vol. 26, p. 132.
7 C. Geertz, *The Interpretation of Cultures*, New York, Basic Books, 1973, p. 433.
8 *Ibid.*, p. 448.
9 *Ibid.*, p. 436.
10 *Ibid.*, p. 449.
11 *Ibid.*, p. 448.
12 *Ibid.*, p. 452.
13 I hesitate to call this view *empirical* only because doing so risks importing the baggage of a particular epistemological tradition that has often been associated (incorrectly) with positivism.
14 *Ibid.*, p. 437.
15 *Ibid.*
16 *Ibid.*
17 *Ibid.*, p. 5.
18 *Ibid.*, p. 6.
19 My emphasis.
20 G. Ryle, "The Thinking of Thoughts," *Collected Papers, Volume II: Collected Essays, 1929–1968*, London, Hutchinson, 1971, p. 487.
21 See G. Ryle, *The Concept of Mind*, London, Hutchinson, 1949.
22 Geertz, *op. cit.*, p. 7.
23 As, for example, in V. Valeri, "On Anthropology as Interpretive Quest," *Current Anthropology*, 1987, vol. 28, pp. 355–357. Of the Balinese cockfight, Geertz writes that it is "fundamentally a dramatization of status concerns" (Geertz, *op. cit.*, p. 437).
24 *Ibid.*, p. 17.
25 *Ibid.*, p. 44.
26 On these matters, it is the work of Winch rather than that of Wittgenstein that is usually appealed to. Speaking of action, Winch asserts that, to the extent that it is symbolic, "it goes together with certain other actions in the sense that it *commits* the agent to behaving in one way rather than another in the future" (P. Winch, *The Idea of a Social Science and its Relation to Philosophy*, London, Routledge and Kegan Paul, 1958, p. 50). This is only possible,

argues Winch, to the extent that "my present act is the *application of a rule*" (*ibid.*). The argument is meant to track the way things work in the Private Language Argument. That is, the very notion of an act only takes place in a social context in which shared rules are constitutive of acts as acts. For the purposes of my argument we can sidestep the question of the veracity of the Private Language Argument itself. However, I think McGinn is on the mark in his attack on the community thesis of the Private Language Argument as it is traditionally conceived. (See C. McGinn, *Wittgenstein on Meaning*, Oxford, Blackwell, 1984, pp. 194–200.) Winch aspires to give an account of action that does not rely on intention. In the case of language, the first move in the Private Language Argument is to think of an attempt to construct a private language. By its nature this involves, more than tokens (single instances of the use) of words, the constitution of types. That is, the construction of a language has to assume stability of reference across different instances of the use of a particular word. Thus, one might think that the constitution of types "commits us to future use." As such, if one agrees with the Private Language Argument, rules come into play at this point. Winch thinks this analysis can be straightforwardly extended to apply to individual acts as well . . . even to acts of what he calls a "private" nature. Yet, in what sense do such acts "commit us to future use"? Winch illustrates the idea as follows: "if *N* places a slip of paper between the leaves of a book he can be said to be 'using a bookmark' only if he acts with the idea of using the slip to determine where he shall start re-reading" (Winch, *op. cit.*, p. 50). But the problem with this sort of example is that we are not provided with an account of what it is to "act with an idea," and we need one. For the phrase looks naturally like what might be called a "forward-directed intention." Yet, intention is just what we are trying to do without. Part of the problem is that there is a type–token confusion here. For, underlying the notion of the commitment to the future, what Winch is really concerned to bring out is the idea that rule-governed behavior is (at least in principle) repeatable as a whole. But that does not mean that tokens themselves have to have any element of future-directedness to them in order to qualify. Be that as it may, analyzing the meaning of custom *may* have the same feature as language in the sense that the interest is at the level of types, not tokens. But that is not enough to get us to the conclusion that instances of a custom carry a commitment to future instances. True, the *acting* out of custom might carry such a commitment, but actors are just what we want to leave out of the picture *qua* meaning. Yet, without actors, in what sense does custom *itself* carry a commitment to future instances? Stephen Turner asks: "So what does Winch mean by saying that meaningful action is *ipso facto* rule governed? No more than this: In order for an act to mean a particular thing to the agent—to be an act of a particular kind to him—it must be governed by a rule, minimally, the rule that identifies the act as meaning that particular thing and not another, or identifies it as an act of that particular kind and not another" (S. Turner, *Sociological Explanation as Translation*, Cambridge, Cambridge University Press, 1980, p. 12). Bracketing the talk of agents, the proposal comes to this: for an act to have meaning, it must be of a kind; and to be of a kind, it must be subject to at least a rule that differentiates between kinds. On this interpretation there is no longer any requirement of a commitment to the future that is implicit in an act for it to have a sense of a meaning. Rather, it just has to be understood as an instance of a type—just as in the case of language—and the claim is that rules are what achieve this. But having said this, we still need to know in just what sense a custom is to be thought of as rule-*governed* in the first place.

27 Geertz, *op. cit.*, p. 5.

28 P. Guiraud, *Semiology*, London, Routledge and Kegan Paul, 1975, p. 42.

29 I agree with Quine when he distinguishes between behavior that merely fits a rule and behavior that is guided by a rule. However, I think that his condition for rule-governed behavior is too strict: "Behavior fits a rule whenever it conforms to it; whenever the rule truly describes it. But the behavior is not guided by the rule unless the behavior knows and can state it" (W. Quine, "Methodological reflections on current linguistic theory," in D. Davidson and G. Harman (eds), *Semantics of Natural Language*, Dordrecht, Reidel, 1972, p. 442). Still, I think that Quine is right in the sense that the proponent of implicit rules has to provide a way to distinguish between rule-governed phenomena and phenomena that merely fit a rule. I do not see a way to do this unless we have the resources to show how implicit rules are causally implicated in the production of phenomena, as, for example, in the way we do in study of the "implicit" rules of genetic regulation (by way of reduction). But the appeal to "implicit" rules when it comes to the customs of a culture is not to a concept of rules that is nearly that strong in the sense of its implication in the causal nexus. Unlike the case of genetics, we lack a substrate of causal analysis into which we can enlighteningly translate our rule-based talk. (Although we can do so unenlighteningly—by way of a causal reductive account available at the token, rather than the type, level.) Lacking such a substrate we may attempt to restrain our talk to explicit rules. But doing so sharply restricts the domain to which rules can be applied. Nor will transforming implicit rules into explicit rules as an act of theoretical construction help. The mere fact that an implicit rule can be rendered explicit does not distinguish between rules that "govern" and those that do not, except in so far as we can locate such rules in a causal network. John Searle (in *The Construction of Social Reality*, New York, Free Press, 1995) seems to suggest a way to solve this problem. One salient feature of Searle's account is that it does not

force all habitualized behavior to count as meaningless. Searle proposes that background non-intentional capacities "can be causally sensitive to the specific forms of constitutive rules without actually containing any beliefs or desires or representations of those rules" (*ibid.*, p. 141). Consider baseball, and allow that it is the rules of baseball that are constitutive of an activity as baseball. That is not to say that players playing baseball have to have those rules in mind as they play in order for them to be taken to be playing baseball. They may be playing out of habit. Yet their habitual play still counts as baseball. Why? Searle's answer is that they are playing baseball because their behavior has been deliberately organized to fit those rules. On Searle's view the chain of connection can become quite attenuated: he writes of a hypothetical isolated culture that follows the rules of baseball without anyone knowing that he or she is doing so. But that they do so is no accident. Instead, sometime in their culture's past, the rules of baseball were introduced. Now they have been lost but have been replaced by a set of behavioral dispositions that are coincident with the application of the rules of baseball. That they are coincident is no coincidence. The dispositions are organized by way of behavioral rewards and punishments (children get three tries at bat but are then told to sit down) that were developed to fit the rules of baseball. The reinforcers remain long after the rules were lost.

There is much to like about this approach, for it holds the promise of a plausible way in which the reach of constitutive rules can be expanded without threatening to subsume everything. The crucial move that allows for this is the appeal to a causal property. It is not enough that behavioral dispositions or the structure of rewards and punishments coincide with the rules of baseball, for behavior that coincides with rules by accident does not count. Behavior has to conform in virtue of the rules even if the rules bear a long-distance relationship to the behavior itself. Once we make clear this causal feature of Searle's approach we can also see how it could be generalized away from his preference for rules. Whatever the relevant feature that makes behavior meaningful, it can either make behavior meaningful directly or in virtue of Searle's causal mechanism. But there is a dark side to this causal construal—we know from other causal accounts that there is causal and causal. All causal theories fall prey to counter-examples of the wrong kind of causal connection (see Chapter One, Note 14).

30 But are meaning and habit inconsistent? Roughly put, to speak of habit here is to speak of a link from symbol unmediated by meaning—roughly, because we ought not to rule out any meaning in the course of habit. Let us put it this way: if meaning is in play under such circumstances, we need to have an account that differentiates it from the central role that Geertz gives to meaning. In *The Metronomic Society*, Michael Young provides an extraordinarily eloquent defense of the role of habit in society and of culture as a

feature of habit. (See M. Young, *The Metronomic Society*, Cambridge, Mass., Harvard University Press, 1988.) Young argues for a conception of habit that antedates its marriage to a theory of reinforcement and that avoids the pejorative features of that association. For Young, habit and its social feature, culture, exist as a propensity that "generates itself without a stimulus to the contrary" (*ibid.*, p. 78). But, irrespective of the status of reinforcement theory, what advantages does habit accord us? Young argues that there are four: habit allows us to increase the skill with which we perform actions; it diminishes fatigue; it allows us to attend to the unforeseen; and it economizes on memory (*ibid.*, pp. 82–85). The details of this list do not really matter much—what does matter is that, so long as there are some advantages, we have a counter-consideration to the implicit prejudice that more is better in matters of meaning. This is especially so in the extension of an argument from individual habit to the social level, where Young argues for a view of custom as the grand regulator of individual habit (*ibid.*, pp. 95–115). See also, although it is much less clear, Pierre Bourdeau, *Outline of a Theory of Practice*, Cambridge, Cambridge University Press, 1977.

31 See also Anthony Cohen, in *The Symbolic Construction of Community*, New York, Methuen, 1985. Cohen provides a lucid defense for a theory of culture in which the existence of a shared system of symbols is central to the community *without a shared sense of meaning for those symbols*. "The reality of community in people's experience . . . inheres in their attachment or commitment to a common body of symbols," (*ibid.*, p. 16) and, allowing for a variety of associative meanings to be attached to those symbols, facilitates rather than hinders this process by removing the need for communities to have unanimity of meanings for their symbols.

32 It should be noted that such an approach sidesteps a line of criticism of a rule-based account that can be developed from Quinean arguments for the indeterminacy of translation. See C. Hookaway, "Indeterminacy and Interpretation," in C. Hookaway and P. Pettit (eds), *Action and Interpretation*, Cambridge, Cambridge University Press, 1978, pp. 17–42; and P. Roth, *Meaning and Method in the Social Sciences*, Ithaca, Cornell University Press, 1987.

33 R. Keesing, "Anthropology as Interpretative Quest," *Current Anthropology*, 1987, vol. 28, p. 357.

34 Geertz, *op. cit.*, p. 12.

35 *Ibid.*, p. 14.

36 *Ibid.*, p. 145.

37 Of course, this is not to say that they are made up of whole cloth—any more than the theoretical constructs of any scientific theory. The analogy is apt. Our scientific constructs can meld and merge features of the phenomenal world in ways that have no direct counterpart, driven as they are by the dictates of what makes for

good scientific theories. (For more, see N. Cartwright, *How the Laws of Physics Lie*, Oxford, Clarendon Press, 1983.)

38 Gilbert Lewis develops a similar connection in *Day of the Shining Red*, Cambridge, Cambridge University Press, 1980. However, in the end, Lewis embraces a cognitivist approach that I reject. Although Lewis argues against grounding the meaning of an instance of a ritual on either particular actor or audience states of mind, he is concerned about the problem of placing constraints on interpretation (*ibid.*, pp. 26 and 221). His solution is to argue against accepting a claim that a ritual (in general) has a particular meaning like grief if no one ever felt grief (*ibid.*, p. 27). In the end, for Lewis, there is no meaning without people, and it is present not in things but in people's minds (*ibid.*, pp. 221 and 222). But, as we will see, the problem with accepting this view as is comes out precisely in the aesthetic parallel.

39 W. Wimsatt and M. Beardsley, "The Intentional Fallacy," in W. Wimsatt, *The Verbal Icon*, Lexington, University of Kentucky Press, 1954, p. 5.

40 However, this is not to say that persons play no role in controlling the prescriptive constraints on culture.

41 D. Meinig, "Symbolic Landscapes," in D. Meinig and J. Jackson, *The Interpretation of Ordinary Landscapes*, New York, Oxford University Press, 1979, pp. 164–191.

42 *Ibid.*, p. 174.

43 *Ibid.*

44 *Ibid.*, p. 175.

45 *Ibid.*, p. 180.

46 A. Rapoport, "Vernacular Architecture and the Cultural Determinants of Form," in A. King (ed.), *Buildings and Society*, London, Routledge and Kegan Paul, 1980, p. 289.

47 *Ibid.*

48 H. Tomlinson, "The Nineteenth-Century English Prison," in A. King, *op. cit.*, p. 111.

49 J. Frykman and O. Löfgren, *Culture Builders*, New Brunswick, Rutgers University Press, 1987, p. 149.

50 J. Gillis, "Ritualization of Middle-Class Family Life in Nineteenth Century Britain," *International Journal of Politics, Culture and Society*, 1989, vol. 3, p. 221.

51 Frykman and Löfgren, *op. cit.*, p. 149.

52 Gillis, *op. cit.*, p. 221.

53 I am assuming, here, that considerations of reasonableness may fall short of considerations of rationality under some circumstances. Where considerations of rationality can be appropriately appealed to, I take it that the distinction between what is and the as if begins to blur.

54 G. Wright, *Moralism and the Model Home*, Chicago, University of Chicago Press, 1980; and *Building the Dream*, Cambridge, Mass., MIT Press, 1981.

55 *Ibid.*, p. 10.
56 *Ibid.*, p. 12.
57 *Ibid.*.
58 *Ibid.*, p. 15.
59 What follows comes primarily from Wright, *Building the Dream.*
60 Wright, *Moralism and the Model Home*, p. 19.
61 L. Davidoff, J. L'Esperance and H. Newby, "Landscape with Figures: Home and Community in English Society," in A. Oakley and J. Mitchell (eds), *The Rights and Wrongs of Women*, Harmondsworth, Penguin, 1976, p. 149.
62 See A. Batteau, *The Invention of Appalachia*, Tucson, University of Arizona Press, 1990.
63 Wright, *Building the Dream*, pp. 100–102.
64 Wright, *Moralism and the Model Home*, pp. 26–29.
65 *Ibid.*, p. 28.
66 *Ibid.*, p. 27.
67 Wright, *Building the Dream*, p. 112.
68 Palliser and Palliser & Co., Architects, "Introductory," in *Palliser's New Cottage Homes and Details*, New York, Palliser and Palliser & Co., 1887.
69 Wright, *Building the Dream*, p. 113.
70 *Ibid.*, pp. 102–103.
71 Wright, *Moralism and the Model Home*, pp. 18 and 97.
72 *Ibid.*, p. 10.
73 *Ibid.*, p. 13.
74 Wright, *Building the Dream*, p. 109.
75 *Ibid.*, p. 145.
76 *Ibid.*, p. 83.
77 *Ibid.*, pp. 82 and 102–103; and Wright, *Moralism and the Model Home*, pp. 18–19 and 43.
78 *Ibid.*, pp. 250–251.
79 Wright, *Building the Dream*, p. 171.
80 Wright, *Moralism and the Model Home*, p. 121.
81 Wright, *Building the Dream*, pp. 161–162.
82 *Ibid.*, p. 119.
83 *Ibid.*, p. 160.
84 *Ibid.*, p. 159.
85 On the replacement of the parlor by the kitchen as the central focus of the house, see *ibid.*, p. 169. However, that shift did not signal a return to the home as a center of production rather than consumption.
86 *Ibid.*, p. 161.
87 Wright, *Moralism and the Model Home*, p. 108.
88 *Ibid.*, pp. 239 and 234.
89 *Ibid.*, pp. 208, 129.
90 *Ibid.*, p. 293.
91 Even as Wright's account demonstrates, the distinction between construction and dissemination is a convenience of theory. It is not

as if this distinction can be made in action, let alone be made temporally. The pattern builders were both constructing meaning and disseminating it at the same time.

92 See, for example, R. Williams, *The Country and the City*, New York, Oxford University Press, 1973.

93 As in, for example, D. James, *The City as Text*, New York, Cambridge University Press, 1990.

94 Wright, *Building the Dream*, pp. 82 and 102; and *Moralism and the Model Home*, pp. 12 and 43.

95 Geertz, *op. cit.*, p. 5.

96 Wright, *Moralism and the Model Home*, p. 12.

6 CONCLUDING WORRIES

1 J. Appleby, L. Hunt and M. Jacob, *Telling the Truth about History*, New York, Norton, 1994, p. 29.

2 *Ibid.*

3 *Ibid.*, p. 283.

4 J. Kloppenberg, "Objectivity and Historicism: A Century of American Historical Writing," *American Historical Review*, 1989, vol. 94, pp. 1011–1030.

5 Most recently, in Michael Roth's performative alternative. (See M. Roth, "Performing History: Modernist Contextualism in Carl Schorske's *Fin-de-Siècle Vienna*," *American Historical Review*, 1984, vol. 99, pp. 729–745.)

6 C. Peirce, *Reasoning and the Logic of Things: the Cambridge conferences lectures of 1898*, Cambridge, Mass., Harvard University Press, 1992, p. 177.

7 Recently Hoopes has suggested that those who do take pragmatism seriously make a mistake in failing to distinguish between Peirce's views and the views of those who followed him in the tradition. (See J. Hoopes, "Objectivity *and* Relativism Affirmed: Historical Knowledge and the Philosophy of Charles S. Peirce," *American Historical Review*, 1993, vol. 98, pp. 1545–1555.) Far from seeing Peirce as a traditional objectivist, as the above quotation would indicate, Hoopes urges us to think of Peirce as holding to a middle ground in which "Objective knowledge is relative because objective reality comes into being in its fullest relation to thought" (*ibid.*, p. 1551). Made as a general claim, this is a position that cries out for some supporting argument, and all Hoopes does is to defer to some of Peirce's more mysterious claims about "thirdness" and the alleged identity between thought and the world (*ibid.*). In fact I think that the notion that reality "comes into being" via thought has to be understood in relation to what is for us today a non-standard conception of the term "real." For Peirce, the debate is between nominalism and realism. The issue is what the status of the property of, for example, whiteness is, as opposed to instances

of it. (For a very good discussion of this, see C. Hookaway, *Peirce*, London, Routledge and Kegan Paul, 1985, pp. 37–40.) On the vagaries of interpreting pragmatism and verificationism, see M. Johnston, "Objectivity Reconfigured: Pragmatism without Verificationism," in J. Haldane and C. Wright, *Reality, Representation and Projection*, New York, Oxford University Press, 1993, pp. 85–130.

8 Kloppenberg, *op. cit.*, p. 1018. My editorial clarification comes from p. 1017.

9 This view is also found in Newton: "in experimental philosophy we are to look upon propositions collected by general induction from phenomena as accurately or very nearly true, notwithstanding any contrary hypothesis that may be imagined, till such time as other phenomena occur, by which they may be either made more accurate, or liable to exceptions" (I. Newton, *Mathematical Principles of Natural Philosophy*, trans. A. Motte, London, 1729; repr. London, Dawes, 1968, p. 204).

10 Although in Kloppenberg, at least, there is a much more ontological strain of argument when he writes of the "discredited dualisms of both positivism and idealism" and the "indeterminacy of truth and the historicity of reason" (Kloppenberg, *op. cit.*, p. 1030), against which he poses a "Jamesean theory of truth." Here I will do no more than list some of the challenges that such a position faces:

1 If positivism and idealism constitute a dualism, it is a different dualism from objectivism and relativism.
2 The relevance of the claim of a thesis about the indeterminacy of truth needs to defended as opposed to a thesis about confirmation theory.
3 If reason has a historicity, the challenge is to show how this historicity affects both descriptive and prescriptive standards of justification.
4 Finally, it is by no means clear that a full blown "Jamesean theory of truth" can accommodate the elements of realism that Kloppenberg endorses. (For more on this last point, see Hoopes, *op. cit.*)

11 There is another objection that is sometimes confused with this objection: namely, that the argument as developed implicitly assumes that we can treat the notion of data as straightforward. Yet what counts as a datum is purely interest-relative. Data is what we constitute as such, and as such data will be as variable as our interests. This objection attacks the use of the data of practice by allowing that, for any set of data, there will always be more data available that is different in nature—as our interests vary. But this is not a very compelling line of argument, at least as it stands. The interest-relativity of data only establishes that not all data need be relevant to a particular interest. This would be a problem if opponents holding incompatible theories counted as having different

interests and hence differently relevant data—but that is just what Objection 1 asserts.

12 A philosophy of science can be descriptive or prescriptive in nature. That distinction would seem to map easily on to a practice-driven view of philosophy of science versus one that treats theory as independent of practice. For, if a theory of science is to be based on scientific practice, how can it be anything other than descriptive? And if theory is to prescribe, then it must draw on resources that are independent of the evidence of practice itself. But this mapping is misleading. For prescriptivism can come in at least two different guises. One can approach the practice of science with an extra-scientific theory about science in hand that dictates how science ought to be. On the other hand, one can eschew such theoretical commitments and still maintain a prescriptive stance toward science. Here the prescriptivism will just be about the practice of science itself. It will dictate favoring some practices over others. Still, what could motivate this second form of prescriptivism except some prior and independent theoretical commitments about the (prescriptive) nature of science? In fact lots of things could play this role with (intrascientific) pragmatics being the most obvious. It is enough that we favor some practices over others because they yield results that are more useful. But does that not in itself betray a theoretical commitment? It may; however, here we would be still be proceeding from practice to (extrascientific) theory and not the other way around.

13 We saw a variant of this argument in Chapter Three where it arose out of Kuhnian considerations. But here we see it has a much broader foundation.

14 Of course, that is just what we assume when we use an instrument to make readings that verify or falsify a theory. All sorts of theoretical commitments underlie our interpretation of what we are measuring, but that does not matter as long as they are neutral between competing theories.

15 For more, see E. Nagel, *The Structure of Science*, New York, Harcourt, Brace and World, 1961.

16 But note that this is a stronger condition than Quine would allow. (See W. Quine, "On Empirically Equivalent Systems of the World," *Erkenntnis*, 1975, vol. 9, pp. 313–328.)

17 For more, see M. Bunzl, "Real World Epistemic Underdetermination," forthcoming.

18 Namely, modularity at the local level. For more, see *ibid.*

BIBLIOGRAPHY

Albertson, E., Abraham, D. and Murphy, M. (1989), "Interview with Joan Scott," *Radical History Review*, 45: 41–59.

Amin, S. (1987), "Approver's Testimony, Judicial Discourse: The Case of Chauri Chaura," *Subaltern Studies*, 5: 166–202.

Appleby, J. (1989), "One Good Turn Deserves Another: Moving Beyond the Linguistic; A Response to David Harlan," *American Historical Review*, 94: 1326–1332.

Appleby, J., Hunt, L. and Jacob, M. (1994), *Telling the Truth about History*, New York, Norton.

Arnold, D. (1982), "Rebellious Hillmen: The Gudem-Rampa Risings 1839–1924," *Subaltern Studies*, 1: 88–142.

Batteau, A. (1990), *The Invention of Appalachia*, Tucson, University of Arizona Press.

Beard, C. (1934), *The Nature of the Social Sciences*, New York, Scribner.

—— (1934), "Written History as an Act of Faith," *American Historical Review*, 39: 219–231.

Becker, C. (1932), "Everyman his Own Historian," *American Historical Review*, 37: 221–236.

Berkhoffer, R. Jr. (1985), *Beyond the Great Story*, Cambridge, Mass., Harvard University Press.

Bhadra, G. (1985), "Four Rebels of Eighteen-Fifty-Seven," *Subaltern Studies*, 4: 229–275.

Bourdeau, P. (1977), *Outline of a Theory of Practice*, Cambridge, Cambridge University Press.

Boyd, R. (1983), "The Current Status of Scientific Realism," *Erkenntnis*, 19: 45–90.

Bradie, M. (1983), "Criteria for Event Identity," *Philosophical Research Archives*, 9: 29–78.

Brand, M. (1976), *The Nature of Causation*, Urbana, University of Illinois Press.

Brand, M. and Walton, D. (eds) (1976), *Action Theory*, Dordrecht, Reidel Publishing.

Broszat, M. and Friedländer, A. (1988), "A Controversy about the Historicization of National Socialism," *New German Critique*, 44: 85–126.
Bunzl, M. (forthcoming), "Real World Epistemic Underdetermination."
Butterfield, H. (1951), *The Whig Interpretation of History*, New York, Scribner.
Cartwright, N. (1983), *How the Laws of Physics Lie*, Oxford, Clarendon Press.
Chakrabarty, D. (1984), "Trade Unions in a Hierarchical Culture: The Jute Workers of Calcutta, 1920–50," *Subaltern Studies*, 3: 116–152.
Cohen, A. (1985), *The Symbolic Construction of Community*, New York, Methuen.
Collingwood, R. (1986), *The Idea of History*, London, Oxford University Press.
Culler, J. (1982), *On Deconstruction*, Ithaca, Cornell University Press.
Danto, A. (1985), *Narration and Knowledge*, New York, Columbia University Press.
Daston, L. and Galison, P. (1982), "The Image of Objectivity," *Representations*, 40: 81–128.
Davidoff, L., L'Esperance, J. and Newby, H. (1976), "Landscape with Figures: Home and Community in English Society," in A. Oakley and J. Mitchell (eds), *The Rights and Wrongs of Women*, Harmondsworth, Penguin, pp. 139–175.
Davidson, D. (1970), "Mental Events," in L. Foster and J. Swanson (eds), *Experience and Theory*, Amherst, University of Massachusetts Press, pp. 79–101.
Davidson, D. and Harman, G. (eds) (1972), *Semantics of Natural Language*, Dordrecht, Reidel Publishing.
Davis, N. (1987), "Women on Top," *Society and Culture in Early Modern France*, Cambridge, Polity Press, pp. 124–151.
Derrida, J. (1976), *Of Grammatology*, Baltimore, Johns Hopkins University Press.
—— (1978), *Writing and Difference*, Chicago, University of Chicago Press.
—— (1981), *Positions*, Chicago, University of Chicago Press.
Devitt, M. (1984), *Realism and Truth*, Princeton, Princeton University Press.
—— (1991), "Aberrations of the Realism Debate," *Philosophical Studies*, 61: 43–63.
Dreyfus, H. and Rabinow, P. (1983), *Michel Foucault, Beyond Structuralism and Hermeneutics*, Chicago, University of Chicago Press.
Duberman, M., Vicinus, M. and Chauncey, M. Jr. (eds) (1989), *Hidden from History: Reclaiming the Gay and Lesbian Past*, New York, Signet.
Ellis, J. (1989), *Against Deconstruction*, Princeton, Princeton University Press.
Field, H. (1982), "Realism and Relativism," *Journal of Philosophy*, 79: 553–567.
Fine, A. (1984), "And Not Anti-Realism Either," *Noûs*, 18: 51–65.

—— (1984), "The Natural Ontological Attitude," in J. Leplin (ed.), *Scientific Realism*, Berkeley, University of California Press, pp. 83–107.
—— (1986), *The Shaky Game: Einstein, Reality and the Quantum Theory*, Chicago, Chicago University Press.
—— (1986), "Unnatural Attitudes: Realist and Instrumentalist Attachments to Science," *Mind*, 95: 149–177.
Foster, L. and Swanson, J. (eds) (1970), *Experience and Theory*, Amherst, University of Massachusetts Press.
Foucault, M. (1972), *The Archaeology of Knowledge*, New York, Pantheon.
—— (1979), *Discipline and Punish*, New York, Vintage.
—— (1979), *The History of Sexuality, Vol. 1, An Introduction*, London, Allen Lane.
—— (1984), "Nietzsche, Genealogy, History," *The Foucault Reader*, New York, Pantheon, pp. 76–100.
—— (1988), *Politics, Philosophy, Culture: Interviews and Other Writings of Michel Foucault*, New York, Routledge, Chapman and Hall.
Frykman, J. and Löfgren, O. (1987), *Culture Builders*, New Brunswick, Rutgers University Press.
Furet, R. (1982), *In the Workshop of History*, Chicago, University of Chicago Press.
Geertz, C. (1973), *The Interpretation of Cultures*, New York, Basic Books.
Gillis, J. (1985), *For Better, For Worse*, New York, Oxford University Press.
—— (1989), "Ritualization of Middle-Class Family Life in Nineteenth Century Britain," *International Journal of Politics Culture and Society*, 3: 213–236.
Gordon, L. (1990), "Review of *Gender and the Politics of History*," *Signs*, 15: 853–858.
—— (1991), "Comments on *That Noble Dream*," *American Historical Review*, 96: 683–687.
Guha, R. (1982), "Preface," *Subaltern Studies*, 1: vii–viii.
—— (1983), "The Prose of Counter-Insurgency," *Subaltern Studies*, 2: 1–42.
—— (1985), "Forestry and Social Protest in British Kumaun, c.1893–1921," *Subaltern Studies*, 4: 54–100.
—— (1989), "Dominance Without Hegemony and Its Historiography," *Subaltern Studies*, 6: 210–309.
Gutting, G. (1989), *Michel Foucault's Archaeology of Scientific Reason*, Cambridge, Cambridge University Press.
Habermas, J. (1988), *On the Logic of the Social Sciences*, Cambridge, Mass., MIT Press.
Haldane, J. and Wright, C. (eds) (1993), *Reality, Representation and Projection*, New York, Oxford University Press.
Hardiman, D. (1987), "The Bhils and Shahukars of Eastern Gujarat," *Subaltern Studies*, 5: 1–54.
Hempel, C. "The Theoretician's Dilemma: A Study in the Logic of Theory Construction," *Aspects of Scientific Explanation*, New York, Free Press, pp. 173–226.

Henningham, S. (1983), "Quit India in Bihar and the Eastern United Provinces: The Dual Revolt," *Subaltern Studies*, 2: 130–179.

Higham, J. (1985), *History*, New York, Garland.

Hookaway, C. (1985), *Peirce*, London: Routledge and Kegan Paul.

Hoopes, J. (1993), "Objectivity *and* Relativism Affirmed: Historical Knowledge and the Philosophy of Charles S. Peirce," *American Historical Review*, 98: 1545–1555.

Hoy, D. (ed.) (1986), *Foucault, A Critical Reader*, Oxford, Blackwell.

Hull, D. (1975), "Central Subjects and Historical Narratives," *History and Theory*, 14: 253–274.

Iggers, G. (1973), "Introduction" to L. Ranke, *The Theory and Practice of History*, Indianapolis, Bobbs-Merrill.

—— (1983), *The German Conception of History*, Middletown, Wesleyan.

Iggers, G. and Powell, J. (eds) (1990), *Leopold von Ranke and the Shaping of the Historical Discipline*, Syracuse, Syracuse University Press.

James, D. (1990), *The City as Text*, New York, Cambridge University Press.

Johnston, M. (1993), "Objectivity Reconfigured: Pragmatism without Verificationism," in J. Haldane and C. Wright (eds), *Reality, Representation and Projection*, New York, Oxford University Press, pp. 85–130.

Keesing, R. (1987), "Anthropology as Interpretative Quest," *Current Anthropology*, 28: 161–176.

Kim, J. (1976), "Events as Property Exemplifications," in M. Brand and D. Walton (eds), *Action Theory*, Dordrecht, Reidel Publishing, pp. 159–177.

King, A. (ed.) (1980), *Buildings and Society*, London: Routledge and Kegan Paul.

Kirkham, R. (1995), *Theories of Truth*, Cambridge, Bradford Books.

Kloppenberg, J. (1989), "Objectivity and Historicism: A Century of American Historical Writing," *American Historical Review*, 94: 1011–1030.

Kripke, S. (1980), *Naming and Necessity*, Oxford, Blackwell.

Kuhn, T. (1962), *The Structure of Scientific Revolutions*, Chicago, University of Chicago Press.

Leplin, J. (ed.) (1984), *Scientific Realism*, Berkeley, University of California Press.

Lewis, G. (1980), *Day of the Shining Red*, Cambridge: Cambridge University Press.

Lloyd, C. (1983), *The Structures of History*, Oxford, Blackwell.

Lorenz, C. (1994), "Historical Knowledge and Historical Reality: A Plea for 'Internal Realism,'" *History and Theory*, 33: 297–327.

McGinn, C. (1984), *Wittgenstein on Meaning*, Oxford, Blackwell.

Mayer, A. (1988), *Did the Heavens Not Darken?*, New York, Pantheon.

Megill, A. (1985), *Prophets of Extremity*, Berkeley, University of California Press.

—— (1989), "Recounting the Past: 'Description,' Explanation and Narrative in Historiography," *American Historical Review*, 94: 627–653.

—— (1991), "Four Senses of Objectivity," *Annals of Scholarship*, 8: 301–320.
Meinig, D. (1979), "Symbolic Landscapes," in D. Meinig and J. Jackson (eds), *The Interpretation of Ordinary Landscapes*, New York, Oxford University Press, pp. 164–191.
Meinig, D. and Jackson, J. (eds) (1979), *The Interpretation of Ordinary Landscapes*, New York, Oxford University Press.
Merquior, J. (1985), *Foucault*, London, Fontana.
Miller, R. (1989), "In Search of Einstein's Legacy: A Critical Notice of *The Shaky Game: Einstein, Reality and the Quantum Theory*," *Philosophical Review*, 98: 215–238.
Murphey, M. (1994), *Philosophical Foundations of Historical Knowledge*, Albany, SUNY Press.
Musgrave, M. (1989), "Noa's Ark – Fine for Realism," *The Philosophical Quarterly*, 39: 383–398.
Nagel, E. (1961), *The Structure of Science*, New York: Harcourt, Brace and World.
Nelson, C. and Grossberg, L. (eds) (1988), *Marxism and the Interpretation of Culture*, Urbana, University of Illinois Press.
Newton, I. (1729), *Mathematical Principles of Natural Philosophy*, trans. A. Motte, London; repr. London, Dawes, 1968.
Norris, C. (1985), *The Content of Faculties*, London, Methuen.
Novick, P. (1988), *That Noble Dream*, Cambridge, Cambridge University Press.
Oakley, A. and Mitchell, J. (eds) (1976), *The Rights and Wrongs of Women*, Harmondsworth, Penguin.
O'Hanlon, R. (1988), "Recovering the Subject Subaltern Studies and Histories of Resistance in Colonial South Asia," *Modern Asia Studies*, 22: 189–224.
O'Hanlon, R. and Washbrook, D. (1992), "After Orientalism: Culture, Criticism and Politics in the Third World," *Comparative Studies in Society and History*, 34: 141–167.
Ortner, S. (1984), "Theory in Anthropology since the Sixties," *Comparative Studies in Society and History*, 26: 126–166.
Palliser and Palliser & Co., Architects (1987), "Introductory," in *Palliser's New Cottage Homes and Details*, New York, Palliser and Palliser & Co., Architects.
Palmer, B. (1990), *Descent into Discourse: The Reification of Language and the Writing of Social History*, Philadelphia, Temple.
Pandey, G. (1982), "Peasant Revolt and Indian Nationalism: The Peasant Movement in Awadh, 1919–1922," *Subaltern Studies*, 1: 143–197.
—— (1983), "Rallying round the Cow: Sectarian Strife in the Bhojpuri Region c.1888–1917," *Subaltern Studies*, 2: 60–129.
—— (1984), " 'Encounters and Calamities': The History of a North Indian *Qasba* in the Nineteenth Century," *Subaltern Studies*, 3: 231–270.
—— (1994), "The Prose of Otherness," *Subaltern Studies*, 8: 188–221.

Peirce, C. (1992), *Reasoning and the Logic of Things: The Cambridge Conferences Lectures of 1898*, Cambridge, Mass., Harvard University Press.

Poovey, M. (1988), *Uneven Developments*, Chicago, University of Chicago Press.

—— (1988), "Feminism and Deconstruction," *Feminist Studies*, 14: 51–65.

Prado, C. (1995), *Starting with Foucault: An Introduction to Genealogy*, Boulder, Westview Press.

Prakash, G. (1992), "Can the 'Subaltern' Ride? A Reply to O'Hanlon and Washbrook," *Comparative Studies in Society and History*, 34: 168–184.

—— (1994), "Subaltern Studies as Postcolonial Criticism," *American Historical Review*, 99: 1475–1490.

Priest, G. (1994), "Derrida and Self-Reference," *Australasian Journal of Philosophy*, 72: 103–111.

Putnam, H. (1975), "The Meaning of 'Meaning,'" *Mind, Language and Reality, Philosophical Papers, Volume 2*, Cambridge, Cambridge University Press, pp. 215–271.

—— (1981), *Reason, Truth and History*, Cambridge, Cambridge University Press.

—— (1988), *Representation and Reality*, Cambridge, Bradford Books.

—— (1990), *Realism with a Human Face*, Cambridge, Mass., Harvard University Press.

Quine, W. (1960), *Word and Object*, Cambridge, Mass., MIT Press.

—— (1972), "Methodological reflections on current linguistic theory," in D. Davidson and G. Harman (eds), *Semantics of Natural Language*, Dordrecht, Reidel, pp. 442–454.

—— (1975), "On Empirically Equivalent Systems of the World," *Erkenntnis*, 9: 313–328.

Ranke, L. (1973), *The Theory and Practice of History*, Indianapolis, Bobbs-Merrill.

Rapoport, A. (1980), "Vernacular Architecture and the Cultural Determinants of Form," in A. King (ed.), *Buildings and Society*, London: Routledge and Kegan Paul, pp. 283–305.

Rorty, R. (1979), *Philosophy and the Mirror of Nature*, Princeton, Princeton University Press.

—— (1986), "Foucault and Epistemology," in D. Hoy (ed.), *Foucault, A Critical Reader*, Oxford, Blackwell, pp. 41–49.

Ross, D. (1990), "On Misunderstanding of Ranke and the Origins of the Historical Profession in America," in G. Iggers and J. Powell (eds), *Leopold von Ranke and the Shaping of the Historical Discipline*, Syracuse, Syracuse University Press, pp. 154–169.

Roth, M. (1982), "Performing History: Modernist Contextualism in Carl Schorske's *Fin-de-Siècle Vienna*," *American Historical Review*, 99: 729–745.

Roth, P. (1987), *Meaning and Method in the Social Sciences*, Ithaca, Cornell University Press.

Ryle, G. (1949), *The Concept of Mind*, London, Hutchinson.

—— (1971), "The Thinking of Thoughts," *Collected Papers, Volume II: Collected Essays*, London, Hutchinson, pp. 480–496.
Schleunes, K. (1970), *The Twisted Road to Auschwitz*, Urbana, University of Illinois Press.
Scott, J. (1988), *Gender and the Politics of History*, New York, Columbia University Press.
—— (1990), "Response to Gordon," *Signs*, 15: 859–869.
—— (1991), "The Evidence of Experience," *Critical Inquiry*, 17: 773–797.
—— (1996), *Only Paradoxes to Offer*, Cambridge, Mass., Harvard University Press.
Searle, J. (1985), *The Construction of Social Reality*, New York, Free Press.
Shapere, D. (1982), "Reason, Reference and the Quest for Knowledge," *Philosophy of Science*, 49: 1–23.
Shiner, L. (1982), "Reading Foucault: Anti-Method and the Genealogy of Power-Knowledge," *History and Theory*, 21: 382–398.
Snyder, P. (ed.) (1958), *Detachment and the Writing of History: Essays and Letters of Carl Becker*, Ithaca, Cornell University Press.
Sober, E. (1984), *The Nature of Explanation*, Cambridge, Bradford Books.
Spivak, G. (1985), "Subaltern Studies: Deconstructing Historiography," *Subaltern Studies*, 4: 330–363.
—— (1988), "Can the Subaltern Speak?," in C. Nelson and L. Grossberg (eds), *Marxism and the Interpretation of Culture*, Urbana, University of Illinois Press, pp. 271–313.
Stedman Jones, G. (1983), *Languages of Class: Studies in English Working Class History, 1832–1982*, Cambridge, Cambridge University Press.
Stein, H. (1989), "Yes, but . . . Some Skeptical Remarks on Realism and Anti-Realism," *Dialectica*, 43: 47–65.
Strout, C. (1958), *The Pragmatic Revolt in American History: Carl Becker and Charles Beard*, New Haven, Yale University Press.
Tomlinson, H. (1980), "The Nineteenth-Century English Prison," in A. King (ed.), *Buildings and Society*, London, Routledge and Kegan Paul, pp. 94–119.
Turner, S. (1980), *Sociological Explanation as Translation*, Cambridge, Cambridge University Press.
Twigg, G. (1984), *The Black Death: A Biological Reappraisal*, London, Batsford.
Valeri, V. (1987), "On Anthropology as Interpretive Quest," *Current Anthropology*, 28: 335–357.
van Fraassen, B. (1980), *The Scientific Image*, Oxford, Clarendon Press.
Veyne, P. (1971), *Comment on écrit l'histoire, suivi de Foucault révolutionne l'histoire*, Paris, Éditions du Seuil.
Weber, M. (1964), *The Theory of Social and Economic Organization*, New York, Free Press.
Weeks, J. (1982), "Foucault for Historians," *History Workshop*, 14: 106–119.
—— (1989), *Sex, Politics and Society*, London, Longman.
White, H. (1978), *Tropics of Discourse*, Baltimore, Johns Hopkins University Press.